Appleton & Lange's Review of PHYSIOLOGY

Appleton & Lange's Review of

PHYSIOLOGY

David G. Penney, PhD
Professor of Physiology
Wayne State University, School of Medicine
Detroit, Michigan

with

Andreas Carl, MD, PhD
Department of Pathology
School of Medicine
University of Colorado Health Sciences Center
Denver, Colorado

Appleton & Lange
Stamford, Connecticut

98 99 00 01 02 / 10 9 8 7 6 5 4 3 2 1

Prentice Hall International (UK) Limited, *London*
Prentice Hall of Australia Pty. Limited, *Sydney*
Prentice Hall Canada, Inc., *Toronto*
Prentice Hall Hispanoamericana, S.A., *Mexico*
Prentice Hall of India Private Limited, *New Delhi*
Prentice Hall of Japan, Inc., *Tokyo*
Simon & Schuster Asia Pte. Ltd., *Singapore*
Editora Prentice Hall do Brasil Ltda., *Rio de Janeiro*
Prentice Hall, *Upper Saddle River, New Jersey*

Library of Congress Cataloging-in-Publication Data

Penney, David G.
Appleton & Lange's review of physiology / David G. Penney.—1st ed.
p. cm.
Includes bibliographical references.
ISBN 0–8385–0274–1 (pbk. : alk. paper)
1. Human physiology—Examinations, questions, etc. I. Title.
[DNLM: 1. Physiology—examinations questions. QT 18.2 P413a 1998]
QP40.P44 1998
612'.0076—dc21
DNLM/DLC 97-33321
for Library of Congress CIP

ISBN 0-8385-0274-1
90000
9 780838 502747

Acquisitions Editor: Marinita Timban
Production Service: Rainbow Graphics, Inc.
Designer: Libby Schmitz

PRINTED IN THE UNITED STATES OF AMERICA

Dedicated to

Joseph Cascarano
Professor of Biology, Emeritus
University of California, Los Angeles

He has been a unique and powerful influence on the careers of his students, fondly remembered by all.

Contents

Preface

As noted in the Table of Contents, this book covers all major areas of human physiology. An attempt has been made to give each section a weight in terms of the number of questions presented, similar to what is found on the current National Medical Licensing Examination on Physiology.

The review of examination questions serves a multitude of learning functions, and acts as an adjunct to the study of a discipline in the traditional manner. It is used as:

1. A diagnostic tool to identify areas of weakness, each of which can then be improved on.
2. A test of the acquisition of facts.
3. A test of problem-solving skills.
4. A predictor of competency.

There is a finite number of good questions that can be written on the body of knowledge in any discipline. Achieving mastery over representatives of that pool of questions means you have mastered the subject material.

Part I of the National Medical Licensing Examination on Physiology has been streamlined in the past few years in that only two types of questions are now used. Thus only the following two types are used in this book:

1. Single answer multiple choice, consisting of a stem followed by 5, but occasionally 4, possible matches.
2. Extended matching, consisting of a stem and up to 20 possible matches.

This book has been prepared with three populations of readers in mind: the student who is attempting to learn physiology for the first time, the student who is reviewing physiology after having completed formal training in the discipline, and the instructor who is assembling an examination. For those beginning the study of physiology, the question–answer format can be an important learning tool. For those who wish to obtain additional information about subjects in physiology, reference to some current texts and electronic media is provided below.

The questions presented here fall into two broad categories: recall of appropriate facts, and problem solving. To work effectively in any discipline, one must know a certain body of facts about the discipline. The facts can then be used in solving problems in the discipline.

It is presumed that in using this book, your goal is not so much to get the right answer as to learn the material. Therefore, don't let a lucky guess deceive you. In taking any examination, don't be misled by extraneous factors.

1. Don't look for a pattern in the answers. Answers are frequently arranged by a randomizing technique.
2. Regard each question as an honest attempt to assess your skills. The student who is looking for trick questions is usually his own worst enemy.
3. Mark your first reasonable answer to a question. Often prolonged pondering leads to changing your first correct answer into one that is incorrect.
4. Don't read into questions material that is not there. For example, where the question states: "On the basis of the following data calculate . . . ," you are not being asked to use all the data but merely to select the relevant data to reach a certain conclusion.
5. Work in the context of the question. In other words, if you are asked to do a calculation and the answers from which you are to select are all to two-place accuracy, don't calculate to the

fourth place. If there is no penalty for guessing, never leave an answer blank.

6. Organize your time. In a timed examination it is usually important to complete the examination. Don't spend 50% of your time on a question worth 1% of the total score. Control that tendency to be compulsive before it is too late. Successful students first survey an examination to gauge its difficulty. Questions that appear too difficult are left on the first pass and taken up later once the easier questions have been answered.

The answers and comments on the learning material in the text follow at the back of each chapter. This arrangement allows you to honestly and deliberately consider each question before looking at the answer. You will undoubtedly find this valuable in internalizing the information and concepts.

Other Reading

1. Berne, R.M., Levy, M.N. *Physiology,* 3rd ed., St. Louis, Mosby, 1993.
2. West, J.B. (ed): *Best and Taylor's Physiological Basis of Medical Practice,* 12th ed., Baltimore, Williams and Wilkins, 1990.
3. Ganong, W.F. *Review of Medical Physiology,* 17th ed., Los Altos, CA, Lange, 1995.
4. Guyton, A.C. *Textbook of Medical Physiology,* 9th ed., Philadelphia, Saunders, 1996.
5. Vander, A.J., Sherman, J.H., Luciano, D.S. *Human Physiology: The Mechanisms of Body Function,* 6th ed., New York, McGraw-Hill, 1994.

Instructional Website

1. Virtual Classroom, Wayne State University School of Medicine, Detroit—
http://www.phypc.med.wayne.edu. Contact webmaster for current security codes, dpenney@cmb.biosci.wayne.edu.

PART I

Introduction

CHAPTER 1

Basic Principles

Questions

DIRECTIONS (Questions 1 through 22): Each of the numbered items or incomplete statements in this section is followed by answers or by completions of the statement. Select the ONE lettered answer or completion that is BEST in each case.

1. Which of the following mathematical relationships is correct?

(A) 1 mm Hg = 1330 dynes/cm^2
(B) 1 calorie = 1330 dynes/cm^2
(C) 1 watt = 10^{-7} joules
(D) 1 erg = 1 dyne cm^{-1}
(E) 1 poise = 2.39×10^{-8} calories

2. Five grams of $CaCl_2$ were added to 200 mL of water. The atomic weight of calcium is 40 g and that of chlorine is 35.5 g. What is the concentration of this solution?

(A) 2.5%
(B) 3.0%
(C) 10 g of $CaCl_2$/L
(D) 15 g of $CaCl_2$/L
(E) 20 g of $CaCl_2$/L

3. The concentration of $CaCl_2$ reported for Question 2 is equivalent to which of the following?

(A) less than 200 mM/L
(B) between 200 and 220 mM/L
(C) between 220 and 230 mM/L
(D) between 230 and 240 mM/L
(E) more than 240 mM/L

4. The concentration of Ca in the solution discussed in Questions 2 and 3 is

(A) less than 225 mEq/L
(B) 225 mEq/L
(C) 300 mEq/L
(D) 440 mEq/L
(E) over 440 mEq/L

5. An aqueous solution contains 0.1 mM of H^+ and 0.1 mM of Cl^- per liter. On the basis of these data, one can conclude that the solution has

(A) a pH of less than 3 and is acid
(B) a pH of less than 3 and is basic
(C) a pH of 4 and is acid
(D) a pH of 4 and is basic
(E) a concentration of OH^- of more than 0.1 mM/L

6. Solution 1 has a pH of 7.4, solution 2 a pOH of 4, solution 3 a hydrogen ion concentration of 10^{-6} M, and solution 4 a hydroxide ion concentration of 10^{-5} M. Arrange the solutions in order of increasing acidity.

(A) 1, 2, 3, 4
(B) 1, 3, 4, 2
(C) 2, 4, 3, 1
(D) 2, 4, 1, 3
(E) 3, 1, 4, 2

7. Intracellular fluid characteristically contains

(A) 95 mEq of Cl^-/L
(B) 80 mEq of Ca^{++}/L
(C) 100 mEq of HCO_3^-/L
(D) 70 mEq of Na^+/L
(E) 155 mEq of K^+/L

8. Which of the following statements concerning intracellular fluid is true?

(A) it contains over 50% of the body water
(B) it has a higher osmotic pressure than extracellular fluid
(C) it has a higher concentration of organic anions than extracellular fluid
(D) A and C are correct
(E) all are correct

9. Which of the following statements concerning the diffusion of water is true?

(A) the net flux of water is from an area of low solute concentration to an area of high solute concentration
(B) requires a semipermeable membrane
(C) is an energy-requiring process
(D) A and C are correct
(E) A and B are correct

10. Nine grams of NaCl are added to 1 liter of water. At 37°C, this solution exerts an osmotic pressure of 7.6 atmospheres. How much osmotic pressure would be exerted by a solution containing 20 g of glucose per liter of water under similar circumstances? The molecular weight of glucose is 180 g and NaCl is 58.5 g.

(A) less than 3 atmos
(B) 5.6 atmos
(C) 7.6 atmos
(D) 10 atmos
(E) more than 15 atmos

11. The following data were obtained from a patient by injecting an indicator and determining its concentration in the blood:

a. total body water (indicator = DHO = heavy water): 42 L
b. extracellular water, fast (indicator = thiocyanate): 11 L
c. extracellular water, slow (indicator = thiocyanate): 19 L
d. plasma water (indicator Evans blue): 3 L
e. blood cell volume (calculated from hematocrit): 5 L

What was the patient's intracellular water volume?

(A) 5 L
(B) 11 L
(C) 16 L
(D) 19 L
(E) 23 L

12. In Question 11, what was the patient's interstitial water volume?

(A) 2 L
(B) 5 L
(C) 8 L
(D) 13 L
(E) 20 L

13. One liter of isotonic saline was injected intravenously into a 70-kg woman. The injection had no effect on capillary hydrostatic pressure. After 15 minutes the

(A) extracellular water would increase by more than 900 mL
(B) interstitial water would increase by more than 900 mL
(C) intracellular water would increase by more than 900 mL
(D) total body water would increase by more than 900 mL
(E) A and D are correct

14. If one month later, 1 liter of plasma was given intravenously to the woman in Question 13, and once again the capillary hydrostatic pressure stayed constant, what would be the effect on fluid volume 15 minutes after the injection?

(A) extracellular water would increase by more than 900 mL
(B) interstitial water would increase by more than 900 mL
(C) plasma volume would be increased by more than 900 mL
(D) total body water would increase by more than 900 mL
(E) A, C, and D are correct

15. Two months later, 1 liter of an isosmolal solution of urea was injected intravenously to the woman in Question 13, and once again the capillary hydrostatic pressure stayed constant, what would be the effect on fluid volume 15 minutes after the injection?

(A) extracellular water would increase by more than 900 mL
(B) interstitial water would increase by more than 900 mL
(C) intracellular water would increase by more than 900 mL
(D) total body water would increase by more than 900 mL
(E) plasma volume would increase by more than 900 mL

16. The urea solution in Question 15 is

(A) hypertonic
(B) hypotonic
(C) isotonic
(D) none of the above
(E) can't determine

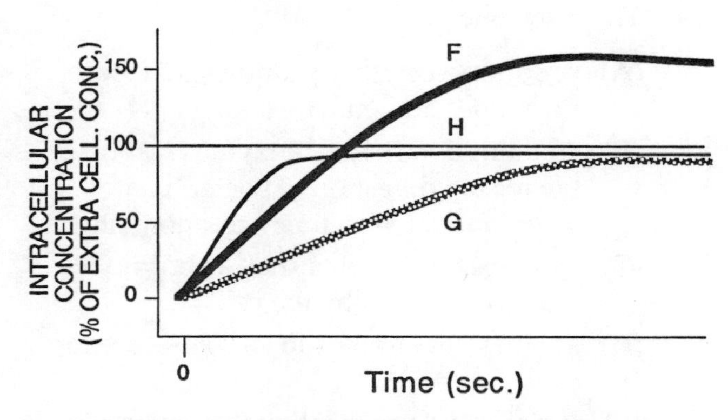

Figure 1–1. Rate of entry of various molecules into a cell.

17. With respect to Figure 1–1, choose the correct statement.

(A) if curves F and G represent molecules that enter only by diffusion through pores, then curve F represents the diffusion of a larger molecule than curve G
(B) if curves G and H represent molecules that enter only through the lipid portion of the cell membrane, then curve H represents a less lipid-soluble molecule than curve G
(C) if curve F represents an ion being moved by facilitated diffusion, then curve G represents the movement of the same ion after ATPase inhibition by ouabain
(D) if curve H represents a molecule being moved by facilitated diffusion, then curve G represents the movement of the same molecule after the addition of a molecule of similar chemical structure
(E) curve F represents the diffusion of a negatively charged particle into the cytoplasm

18. Mitochondria are the major source of energy in the cell because they contain

(A) the glycolytic enzymes
(B) electron transport chain
(C) A and B
(D) Krebs–citric acid cycle enzymes
(E) B and D

19. The lysosome

(A) produces secretory granules that store hormones and enzymes
(B) contains a variety of enzymes responsible for the digestion of bacteria and worn out native cellular components
(C) synthesizes proteins such as hormones that are secreted by the cell
(D) synthesizes steroids in steroid-secreting cells
(E) is a site to detoxification in some cells

20. Agranular endoplasmic reticulum

(A) produces secretory granules that store hormones and enzymes
(B) contains a variety of enzymes responsible for the digestion of bacteria and worn out native cellular components
(C) synthesizes proteins such as hormones that are secreted by the cell
(D) synthesizes steroids in steroid-secreting cells
(E) is a site to detoxification in some cells

21. Ribosome on endoplasmic reticulum

(A) produces secretory granules that store hormones and enzymes
(B) contains a variety of enzymes responsible for the digestion of bacteria and worn out native cellular components
(C) synthesizes proteins such as hormones that are secreted by the cell
(D) synthesizes steroids in steroid-secreting cells
(E) is a site to detoxification in some cells

22. The Golgi complex

(A) produces secretory granules that store hormones and enzymes
(B) contains a variety of enzymes responsible for the digestion of bacteria and worn out native cellular components
(C) synthesizes proteins such as hormones that are secreted by the cell
(D) synthesizes steroids in steroid-secreting cells
(E) is a site to detoxification in some cells

Answers and Explanations

1. **(A)** Pressure is measured in both mm Hg and dynes/cm^2. Answers B through E consist of incorrect mixtures of units (ie, comparing apples and oranges). The calorie is a unit of work, and the dyne/cm^2 is a unit of pressure. The watt is a unit of power, and the joule is a unit of work. The erg is a unit of work, and the dyne/cm is a unit of tension. One dyne cm^{-1} = 1 dyne/cm. The poise is a unit of viscosity and the calorie is a unit of work.

2. **(A)**

   % conc. = wt. (g)/100 mL solution = 5 g/200 mL = 2.5 g/100 mL, or 2.5 %

3. **(C)**

   Molar concentration = (g of X/M.W.)/L solution = (5/111)/.200 = 225 mM/L

4. **(E)**

   Concentration = (molar concentration) × (valence) = (225 mM/L) × (2 mEq/mM) = 450 mEq/L

5. **(C)**

   $pH = -\log (H^+)$ $(H^+) = 0.1$ mM/L = 10^{-4} mol/L = $-\log (10^{+4}) = 4$

   Since the solution has a pH of less than 7, it can be characterized as acid (ie, it has a concentration of H^+, which is greater than 10^{-7} mol/L).

   $$(H^+) \times (OH^-) = 10^{-14}$$
   $$(OH^-) = (10^{-14})/(H^+) + (10^{-14})/(10^{-4}) = 10^{-10} = 0.0000000001 \text{ mol/L}$$

6. **(D)**

   Solution 1: pH = 7.4
   Solution 2: pH = 10; pOH = 4
   Solution 3: pH = 6; $(H^+) = 10^{-6}$ mol/L
   Solution 4: pH = 9; $(OH^-) = 10^{-5}$ mol/L

   The least acid pH is 10 (solution 2). The most acid pH is 6 (solution 3).

7. **(E)** The intracellular environment contains approximately 190 mEq of cations and 190 mEq of anions per liter. The major cation is K^+ (157 mEq/L), and the major anions are phosphate (113 mEq/L) and proteins (74 mEq/L). Cl^-, Ca^{++}, HCO_3^-, and Na^+ are all found in higher concentration in the blood plasma and other intercellular fluids than in the intracellular fluid.

8. **(D)** The intracellular fluid contains about 55% of the body's water and over four times the concentration of protein anions than does either plasma or interstitial fluid.

9. **(A)** Diffusion is a random movement of particles that may occur either through semipermeable membranes or in their absence and does not require a source of energy.

10. **(A)** The osmotic pressure exerted by a solution is directly related to the molar concentration of particles in that solution.

Molar concentration = (g of X/MW)/L solution

NaCl: (9 g/L)/(58.5 g/mol) = 0.15 mol/L
Glucose: (20 g/L)/(180 g/mol) = 0.11 mol/L

Since each particle of NaCl ionizes to two particles (Na^+ and Cl^-) when placed in solution, the NaCl solution contains 0.30 mol of particles per liter. If we assume negligible ionization for glucose, the osmotic pressure of the glucose solution will be:

$$(0.11\ \text{mol/L})/(0.30\ \text{mol/L}) \times (7.6\ \text{atmos}) = 2.8\ \text{atmos}$$

11. **(E)** Intracellular volume = a − c = 42 − 19 = 23 liters. Line b represents that part of the extracellular water that is freely accessible to the indicator. It equals interstitial volume + plasma volume. Line c includes b (fast extracellular volume) + transcellular volume (in the joints and cerebral ventricles) + cartilage and bone water.

12. **(C)** Interstitial volume = b − d = 11 − 3 = 8 liters.

13. **(E)** The solute (Na^+) would be distributed in the plasma and interstitial space. Therefore, the extracellular and total body water would each be increased by about 1 liter and the intracellular volume would remain unchanged. Since the plasma volume represents about 27% of the fast extracellular space, its volume would increase by about 270 mL. The interstitial volume represents 73%, so its volume would increase by about 730 mL.

14. **(E)** The colloids in the plasma do not readily move into the interstitial space, and therefore the plasma volume 15 minutes after the injection will be increased by about 1 liter, and the interstitial and intracellular space will be unchanged.

15. **(D)** Since urea freely passes the capillary and cell membrane barrier, it will be distributed throughout the extracellular and intracellular spaces; the intracellular volume will increase 670 mL if none of the water enters cartilage, bone, or transcellular water.

16. **(C)** A red blood cell placed in isosmolar urea will swell. This is because urea diffuses into the cell and brings water with it. It is the impermeants (particles that do not permeate or move through cell membranes) that determine tonicity; the impermeants plus the permeants determine osmolarity.

17. **(D)** The rate of facilitated diffusion of a molecule will be inversely related to the concentration of other molecules competing for the transport system. The rate of diffusion through pores is inversely related to the size of the molecule. The rate of diffusion through lipid is directly proportional to the degree of lipid solubility. Facilitated diffusion is not an energy-requiring phenomenon. Since the cytoplasmic side of the cell membrane is negatively charged, curve F could represent the diffusion of a positively charged ion into the cell. Negatively charged ions that diffuse through the cell membrane would obtain an intracellular concentration which is less than their extracellular concentration. For example, HCO_3^- is formed in the cell from CO_2 produced by the cell, but has a lower intracellular concentration than its extracellular concentration.

18. **(E)** Mitochondria contain the enzymes responsible for oxidative phosphorylation. The oxidative phosphorylation of 2 mol of pyruvic acid that occurs during the citric acid cycle results in 13 times more energy for the cell than the anaerobic catabolism of 1 mol of glucose during glycolysis.

19. **(B)**

20. **(D, E)**

21. **(C)**

22. **(A)**

PART II

Peripheral Nervous Control

CHAPTER 2

Organization of the Peripheral Nerves

Questions

DIRECTIONS (Questions 23 through 42): Each of the numbered items or incomplete statements in this section is followed by answers or by completions of the statement. Select the ONE lettered answer or completion that is BEST in each case.

23. A disease of unknown etiology has affected the right leg and thigh in such a way that you suspect that the unmyelinated neurons in that part are not conducting impulses. Which of the following test results on the affected part would give you evidence that your hypothesis is correct?

(A) increased period of latency in the withdrawal reflex in response to pain
(B) inability to perceive pain
(C) increased period of latency in the knee jerk
(D) paralysis of skeletal muscle
(E) decreased sensation of dull pain

24. The speed at which a myelinated axon conducts an action potential is directly related to

(A) the diameter of the dendrites
(B) the diameter of the axon
(C) the length of the axon
(D) the amount of axonal branching
(E) the quantity of acetylcholine initiating the action potential

25. An investigator (1) cuts a peripheral nerve, (2) stimulates its central end with a single supermaximal stimulus, and (3) records electrical activity several centimeters distal of the point of stimulation. What sort of electrical activity does he or she find?

(A) minimal, since neurons cannot produce or conduct action potentials if their axons have been severed
(B) a single spike potential
(C) multiple spike potentials, since reverberating circuits are characteristic of most nerves
(D) multiple spike potentials, because unmyelinated neurons conduct more rapidly than myelinated neurons
(E) multiple spike potentials, because myelinated neurons conduct more rapidly than unmyelinated neurons

26. In an experiment, the axon of a motor neuron is stimulated with a single superthreshold stimulus. In this experiment one expects to record

(A) no action potential in that part of the axon central to the site of stimulation
(B) no action potential in that part of the axon peripheral to the site of stimulation
(C) an action potential in all parts of the axon
(D) an action potential in all parts of the axon and cell body
(E) an action potential throughout the motor neuron and in the neurons innervating the motor neuron

27. All of the following nerves contain visceral afferent fibers EXCEPT

(A) dorsal roots of spinal nerves
(B) vagus nerves
(C) optical nerves
(D) phrenic nerves
(E) splanchnic nerves

28. All of the following fibers travel with the vagus nerve EXCEPT

(A) afferent neurons from mechanoreceptors in the heart and pulmonary vessels
(B) afferent neurons from mechanoreceptors and chemoreceptors in the aorta
(C) afferent neurons from mechanoreceptors and chemoreceptors in and near the carotid sinus
(D) preganglionic parasympathetic neurons innervating the bowels
(E) preganglionic parasympathetic neurons innervating the heart

29. Information concerning the intensity of a stimulus is a function of each of the following EXCEPT

(A) the amplitude of the generator potential in a sensory neuron
(B) the amplitude of the action potential in a sensory neuron
(C) the frequency of action potentials in a single neuron
(D) the number of sensory neurons activated
(E) the modality of sensory neurons activated

30. A single threshold stimulus applied to a somatic efferent neuron will result in

(A) a twitch contraction of a single muscle fiber
(B) a twitch contraction of a group of muscle fibers
(C) a tetanic contraction of a single muscle fiber
(D) a tetanic contraction of a group of muscle fibers
(E) an increase in excitability of the fibers that neuron innervates

31. Acetylcholine is a neurotransmitter in many parts of the body. In which of the following is acetylcholine NOT the *transmitter* substance?

(A) the stimulation of smooth muscle by a postganglionic sympathetic neuron
(B) the stimulation of smooth muscle by a postganglionic parasympathetic neuron
(C) the stimulation of a skeletal muscle fiber by a somatic efferent neuron
(D) the stimulation of a postganglionic parasympathetic neuron by a preganglionic parasympathetic neuron
(E) the stimulation of the adrenal medulla by a preganglionic sympathetic neuron

32. The usual reflex response to the electrical stimulation of a pain fiber from the hand is

(A) facilitation of the motor neurons to the ipsilateral extensor muscles for the forearm
(B) inhibition of the motor neurons to the ipsilateral flexor muscles for the forearm
(C) facilitation of the motor neurons to the ipsilateral extensor and inhibition of the motor neurons to the ipsilateral flexor muscles for the forearm
(D) facilitation of the motor neurons to the contralateral flexor muscles for the forearm
(E) facilitation of the motor neurons to the contralateral extensor muscles for the forearm

33. Which of the following reflexes in the adult disappear in the absence of functional connections between the spinal cord and the brain?

(A) vomiting reflex
(B) sweating reflex

(C) withdrawal reflex
(D) erection of the penis
(E) micturition reflex

34. Interneurons

(A) are an essential part of the withdrawal (nociceptive) reflex
(B) are an essential part of the stretch (myotatic) reflex
(C) are an essential part of all reflexes
(D) are always excitatory
(E) are always inhibitory

35. Muscle spindles are specialized receptors for stretch that contain annulospiral endings. All of the following situations would cause the stimulation of the annulospiral endings in the triceps muscle of the arm EXCEPT

(A) contraction of the extrafusal fibers of the triceps
(B) contraction of the intrafusal fibers of the triceps
(C) contraction of antagonistic muscles
(D) relaxation of synergistic muscles
(E) firing of gamma motor neurons to the triceps

36. The Golgi tendon organ differs in function from the annulospiral endings in that the tendon organ

(A) is more sensitive to the distending force
(B) causes a reflex inhibition of ipsilateral, synergistic alpha neurons
(C) causes a reflex inhibition of ipsilateral, antagonistic alpha neurons
(D) causes a reflex inhibition of contralateral alpha neurons
(E) causes a reflex excitation of contralateral alpha neurons

37. All of the following statements about the gamma efferent neuron are true EXCEPT

(A) it is an A group motor neuron with a diameter smaller than that for alpha efferent neurons
(B) it innervates intrafusal fibers
(C) it innervates muscle fibers that stretch the annulospiral endings
(D) it causes reflex stimulation of alpha efferent neurons
(E) it innervates muscle fibers that stretch the Golgi tendon organ

38. During an operation, a patient has several of his lumbar, posterior roots destroyed. After the operation the patient experiences difficulty in standing erect and a decrease in muscle tone. The reason for these symptoms is that the operation

(A) has destroyed some of the somatic efferent neurons to the leg and thigh
(B) has destroyed sympathetic neurons to the leg and thigh
(C) has destroyed parasympathetic neurons to the leg and thigh
(D) has destroyed neurons carrying impulses from the Krause end bulbs
(E) has destroyed neurons from the annulospiral endings

39. As a result of a motor vehicle accident, the midbrain in a patient is severed between the inferior and superior colliculi. This resulted in a marked increase in extensor muscle tone and an extensor muscle hyperreflexia. Both of these symptoms can be relieved by transection of the dorsal roots. What might be responsible for these symptoms?

(A) generalized loss of facilitation
(B) generalized loss of inhibition
(C) decreased stretch of annulospiral endings in muscle
(D) loss of inhibition delivered to gamma efferent neurons
(E) decreased alpha efferent motor activity

40. A normal, healthy person is attempting to lift a heavy load, and his/her muscles "give out." After this experience, there is no apparent problem in neuromuscular function. What is the most likely mechanism responsible for the abrupt cessation of skeletal muscle contraction?

(A) the activation of stretch receptors in the Golgi tendon organ
(B) the activation of stretch receptors in the annulospiral endings
(C) skeletal muscle ischemia
(D) the inactivation of stretch receptors in the Golgi tendon organ
(E) the inactivation of stretch receptors in the annulospiral endings

41. A patient comes to your office complaining of muscle weakness. You administer neostigmine intramuscularly, and the muscle weakness disappears. What is the mechanism of action of neostigmine?

(A) it blocks the action of acetylcholine
(B) it blocks the action of norepinephrine
(C) it interferes with the action of amine oxidase
(D) it interferes with the action of acetylcholine esterase
(E) it interferes with the action of carbonic anhydrase

42. Decentralization is a process in which all the nerves going to and coming from an organ are severed. This separation of an organ from the central nervous system results in

(A) an increased frequency of contraction in the case of the heart
(B) a contraction that is more forceful in the case of the diaphragm
(C) hypertrophy in the case of skeletal muscle
(D) a decreased sensitivity to acetylcholine in the case of the salivary glands
(E) inability to produce concentrated urine in the case of the kidney

Answers and Explanations

23. (E) There are two types of pain fibers: (1) large-diameter, myelinated type A neurons and (2) small-diameter, unmyelinated dorsal root C fibers. The complete absence of the latter would not affect the period of latency. It might, on the other hand, decrease the duration of sensation and eliminate the dull lingering pain. The afferent limb of the knee jerk reflex contains proprioceptor neurons (ie, neurons from stretch receptors). The efferent limb of the reflex contains somatic efferent neurons. Both of these limbs are large-diameter, myelinated type A neurons.

24. (B) The fastest conducting unmyelinated neurons have the largest diameter. Myelinated neurons conduct more rapidly than unmyelinated neurons with a similar axon diameter.

25. (E) Most nerves are collections of myelinated and unmyelinated neurons of different diameters. Since a supermaximal stimulus is, by definition, a signal sufficiently strong to excite all the neurons in the nerve, and since the velocity of conduction is directly related to the diameter of the neuron and the presence of a myelin sheath, it will take different periods of time for each neuron to send its signal from the point of origin to the recording electrode. This results in multiple action potentials arriving at different points in time.

26. (C) An axon can conduct either toward or away from its cell body. In a motor neuron, a stimulus conducted toward the central nervous system is called an *antidromic impulse,* and one moving away from the central nervous system is called *orthodromic.* In the case of a sensory neuron, conduction away from the cord or brain is called antidromic. The significance of an antidromic stimulus is discussed under *axon reflex.* The cell body does not exhibit action potentials. Rather, it produces an excitatory postsynaptic potential. Motor neurons do not cause action potentials in the neurons that innervate them. The synapse between motor and internuncial neurons permits only impulse transmission to the motor neuron.

27. (C) Both afferent and efferent neurons are found side by side in most of the peripheral nerves. Areas where this is not characteristic include the dorsal and ventral roots of the spinal nerves, the gray rami communicantes, and certain cranial nerves (ie, I, II, and VIII). Vagus nerves contain sensory neurons from the lungs, trachea, esophagus, aortic arch, aortic bodies, and most of the abdominal viscera. Phrenic nerves contain afferent neurons from the pericardium.

28. (C) The vagus contains descending preganglionic parasympathetic neurons innervating many organs including the heart and bowels. The cardiopulmonary and aortic bodies send their signals up the vagi. In contrast, receptors from the carotid sinus send their signals up the ninth cranial nerve.

29. **(B)** The relationship between the height of the action potential in a single neuron and the stimulus is an all-or-none relationship. One does not change the character of the action potential by increasing the intensity of stimulation above threshold. On the other hand, the amplitude of the action potential is modified by changes in the environment. Within limits, as one increases the intensity of a stimulus, more sensory neurons will be stimulated, and each will conduct impulses at a progressively greater *frequency*. In addition, neurons of other modalities (ie, pain fibers) will activate when the stimulus intensity is high.

30. **(B)** In skeletal muscle, a single stimulus to one of its motor neurons causes a single contraction in each of the fibers in that particular *motor unit*. A maintained contraction (ie, a tetanic contraction) normally occurs only in response to a series of stimuli. Autonomic neurons frequently act on cardiac and smooth muscle by increasing or decreasing their irritability. The role of the somatic efferent neuron, on the other hand, is not to change excitability but to initiate contraction.

31. **(A)** Postganglionic sympathetic neurons act on smooth muscle cells via the release of norepinephrine.

32. **(E)** One responds reflexly to pain by ipsilateral (same side as pain) flexion and contralateral (opposite side) extension. The antagonistic muscles to these reactions are reflex inhibited.

33. **(A)** Neuron connections essential for the vomiting reflex are located in the cervical cord and the medulla of the brain. The other listed reflexes have sufficient synaptic connections in the cord to permit them to continue functioning after the loss of brain function. These reflexes may be considerably modified, however, since the normally functioning brain acts to inhibit and facilitate reflex action in the cord.

34. **(A)** The withdrawal reflex, like most reflexes in the body, is multisynaptic (ie, involves at least one internuncial neuron). The knee jerk and other stretch reflexes are monosynaptic. Reflexes involving interneurons may be either inhibitory or excitatory.

35. **(A)** The extrafusal fibers are parallel to the annulospiral endings, and therefore their contraction would decrease any stretch of the endings (ie, prevent their stimulation). Synergistic muscles would have a similar action. The intrafusal fibers are in series with the endings, and therefore their contraction would cause the stretch of the endings. Stretching the entire biceps muscle, by the contraction of antagonistic muscles or other means, would also stretch the endings.

36. **(B)** The tendon organ seems to protect the muscle from a reflex contraction that might do damage. It performs this function through a multisynaptic reflex that inhibits motor neurons to the ipsilateral synergistic muscles. A threshold distending force for the annulospiral endings is about 2 g, whereas threshold for the Golgi tendon organ is about 200 g.

37. **(E)** The stimulation of gamma efferent neurons causes a contraction of intrafusal fibers and a stretch of the annulospiral endings in the muscle spindle, which causes a reflex stimulation of alpha efferent neurons, which causes the stimulation of extrafusal fibers and, therefore, an increase in muscle tone.

38. **(E)** The annulospiral endings are stimulated by muscle stretch and produce a reflex stimulation of alpha efferent neurons and, therefore, a contraction of the homonymous muscle. These stretch reflexes help maintain a high level of muscle tone and are an important part of our postural reflex system. Destruction of the dorsal roots does not cause the destruction of motor neurons. The Krause end bulbs are cold receptors. They are not involved in postural reflexes.

39. **(D)** The symptoms reported can result from a loss of inhibition, but the fact that they can be relieved by transection of the sensory roots indicates a more specific mechanism: Increased stimulation of gamma efferent neurons (1) facilitates more contraction of intrafusal muscle fibers, (2) facilitates more stretch of annulospiral endings, (3) facilitates more reflex stimulation of alpha efferent neurons, and (4) brings about increased muscle tone (extrafusal fibers).

40. (A) The receptors in the Golgi tendon organ have a higher threshold for stretch than those in the annulospiral endings and are, therefore, activated only in response to possibly dangerous distention. This causes a reflex inhibition of the homonymous motor neurons to the extrafusal fibers and, therefore, a cessation of contraction. The receptors in the annulospiral endings facilitate further contraction of the extrafusal fibers. This facilitation can be overwhelmed by inhibition initiated by the Golgi tendon organ. An inadequate blood flow is an unlikely mechanism in a short-term contraction in a healthy subject.

41. (D) Acetylcholine esterase is the enzyme that catalyzes the destruction of acetylcholine. By interfering with the body's natural destroyer of ACh you can exaggerate ACh's action. In the disease *myasthenia gravis,* neostigmine causes an increased muscle strength. In a healthy individual, on the other hand, there is no noticeable change in strength in response to neostigmine. Acetylcholine (ACh) initiates skeletal muscle contraction; blocking its action would make the muscles weaker. Norepinephrine does not play an important role in controlling skeletal muscle contraction. Amine oxidase is an enzyme that facilitates the destruction of catecholamines (norepinephrine, etc.). Carbonic anhydrase catalyzes the formation of H_2CO_3 from CO_2 and water.

42. (A) The heart, unlike skeletal muscle, continues to contract in the absence of any connection with the central nervous system. In human beings, a decentralization of the heart, produced by drugs or by surgery, results in the heart rate changing from about 70 to about 110 beats per minute. Apparently, in human beings at rest, the cardiac nerves have an overall effect of decreasing the rate of firing of the cardiac pacemaker. The diaphragm, like all other skeletal muscles, stops contracting when it is decentralized. All skeletal muscle slowly atrophies in response to decentralization. Salivary glands, as well as most other organs, become hypersensitive to acetylcholine after decentralization. Nerves are far less important in the control of the kidney than hormones such as aldosterone, parathormone, and antidiuretic hormone (ADH). ADH facilitates the production of concentrated urine.

CHAPTER 3

Transmembrane Potential

Questions

DIRECTIONS (Questions 43 through 61): Each of the numbered items or incomplete statements in this section is followed by answers or by completions of the statement. Select the ONE lettered answer or completion that is BEST in each case.

43. Under a particular set of experimental conditions, a mammalian axon was found to be permeable to only one ion, K^+. The equilibrium potential for K^+ in this axon was –97 mV. On the basis of these data, we will expect to find

(A) that the K^+ concentration inside the cell is 5 to 10 times that outside the cell when the resting potential is –97 mV
(B) approximately the same concentration of K^+ inside and outside the cell when the resting potential is –97 mV
(C) when the resting potential is –97 mV, an intracellular concentration of negatively charged ions that is more than 0.5% higher than the intracellular concentration of positively charged ions
(D) an increase in the transmembrane potential if we increase the permeability of the membrane to Na^+
(E) a resting transmembrane potential of –97 mV

44. In the upstroke phase of a nerve action potential, there is a movement of the transmembrane potential

(A) toward the equilibrium potential for Na^+
(B) toward the equilibrium potential for K^+
(C) toward the equilibrium potential for Cl^-
(D) toward the equilibrium potential for Ca^{2+}
(E) no change in transmembrane potential

45. Which of the following statements is FALSE? In a resting skeletal muscle cell

(A) the electrical and chemical gradients for Na^+ are in the same direction
(B) the electrical and chemical gradients for Cl^- are in opposite directions
(C) the electrical and chemical gradients for K^+ are in the same direction
(D) K^+ efflux is impeded by the electrical gradient
(E) Na^+ efflux is impeded by the chemical and electrical gradients

46. Select the mechanism most responsible for the production of the resting transmembrane potential in an axon or skeletal muscle fiber.

(A) a pump that extrudes fewer positively charged particles from the cell than it brings into the cell
(B) a pump that extrudes more K^+ than it brings in Na^+
(C) a pump that extrudes more Na^+ than it brings in K^+
(D) a membrane that is more permeable to Na^+ than to K^+
(E) a membrane that is more permeable to Na^+ than to Cl^-

47. The following concentrations are noted inside and outside a cell:

	INSIDE	OUTSIDE
Na^+	10 mM	115 mM
K^+	120 mM	5 mM
Cl-	? mM	120 mM
$Protein^-$	? mM	0 mM

If the cell volume remains constant and the cell membrane is freely permeable to K^+ and Cl^- but impermeable to Na^+ and $protein^-$, what would the concentration of Cl^- be inside the cell?

(A) 5 mM
(B) 10 mM
(C) 40 mM
(D) 110 mM
(E) 130 mM

48. Using the data from Question 47, determine what concentration of protein would be necessary for osmotic balance.

(A) 10 mM
(B) 25 mM
(C) 85 mM
(D) 105 mM
(E) 120 mM

49. If the permeability of a resting skeletal muscle cell to K^+ is increased while the permeability of the cell to Na^+ stays constant

(A) the transmembrane potential would decrease
(B) the cell would become less excitable because of a decrease in the transmembrane potential
(C) the cell would become more excitable because of a decrease in the transmembrane potential
(D) the transmembrane potential would increase
(E) the transmembrane potential would not change

50. In an experiment, an investigator (1) isolates an axon and places it in Ringer's solution, (2) inserts an intracellular electrode through the cell membrane, (3) measures the resting transmembrane potential, (4) stimulates the axon, and (5) measures its action potential (see Fig. 3–1, Record A). The investigator then replaces the Ringer's solution with one of a different electrolyte content and again records the axon's resting transmembrane potential and action potential (see Fig. 3–1, Record B). On the basis of these data how do you think the second solution differs from the Ringer's solution?

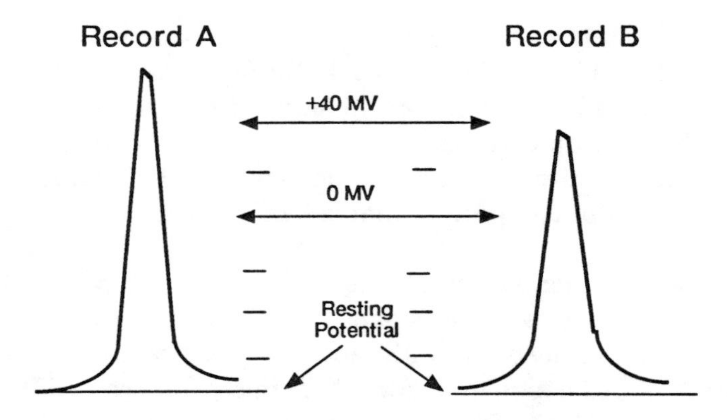

Figure 3–1. Action potentials in an axon.

The second solution contains

(A) a lower concentration of Na^+
(B) a lower concentration of K^+
(C) a higher concentration of Na^+
(D) a higher concentration of K^+
(E) a higher concentration of K^+ and a lower concentration of Na^+

51. A surgeon accidentally removes a patient's four parathyroid glands. The resulting decrease in extracellular Ca^{2+} concentration causes all of the following EXCEPT

(A) decreased membrane stability
(B) increased nerve excitability
(C) increased muscle excitability (tetany)
(D) prolonged ST segment on ECG recordings
(E) impaired blood clotting

52. The following records (Fig. 3–2) were obtained before and after a change in the external environment of the uterus.

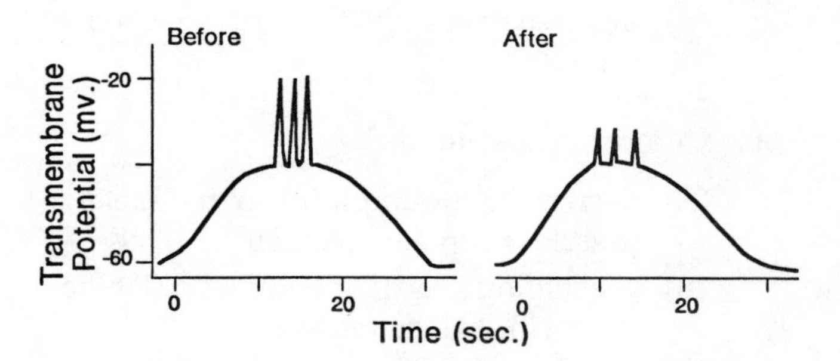

Figure 3–2. Transmembrane potential of uterine cells.

What was the change? The extracellular concentration of

(A) Na^+ was increased
(B) Ca^{++} was increased
(C) K^+ was increased
(D) K^+ was decreased
(E) Ca^{++} was decreased

53. In excitable cells, repolarization is most closely associated with which of the following events?

(A) Na^+ efflux
(B) Na^+ influx
(C) K^+ efflux
(D) K^+ influx
(E) activation of the Na^+/K^+ pump

54. The following action potential (Fig. 3–3) is recorded from a somatic efferent neuron:

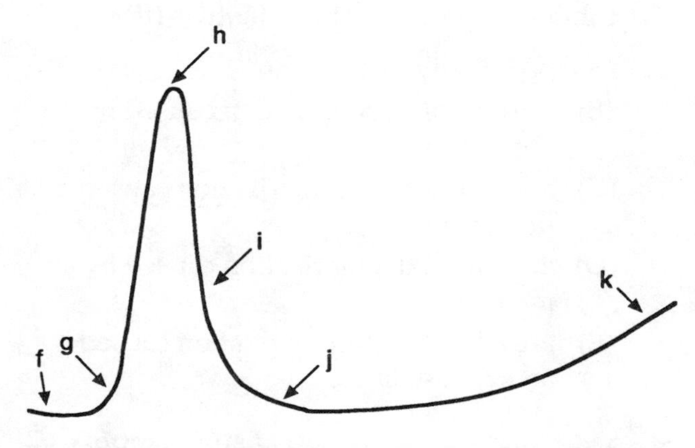

Figure 3–3. Action potential of a somatic efferent neuron.

In the action potential illustrated above

(A) interval g–h is caused by an active transport of Na^+ into the cell
(B) interval g–h is caused by a diffusion of Na^+ into the cell
(C) interval g–h is caused by pinocytosis
(D) interval h–i is caused by active transport of K^+
(E) interval h–i is caused by active transport of Na^+

55. The end-plate potential of skeletal muscle is best characterized as

(A) a local reversal of charge originating at the end plate
(B) a reversal of charge originating at the end plate and propagated throughout the cell
(C) a decrease in the transmembrane potential that is propagated throughout the cell
(D) a local decrease in the transmembrane potential that is caused by an increased permeability to Na^+ and K^+
(E) a local decrease in the transmembrane potential that is associated with little or no increase in Na^+ conductance

56. The excitatory postsynaptic potential (EPSP) differs from the end-plate potential in that the EPSP is

(A) propagated
(B) a reversal of cellular charge
(C) not associated with an increased permeability to Na^+
(D) not decreased by curare
(E) initiated by acetylcholine

57. Inhibitory postsynaptic potentials (IPSPs)

(A) occur only in the brain
(B) are a local depolarization caused by an increase in $g_{Ca^{++}}$
(C) are a local hyperpolarization caused by a decrease in $g_{Ca^{++}}$
(D) are a local depolarization caused by an increase in g_{Cl^-}
(E) are a local hyperpolarization caused by an increase in g_{Cl^-}

58. Miniature end-plate potentials (MEPPs) are found at the postjunctional membrane of the neuromuscular junction. Their amplitude is usually about 0.5 mV. MEPPs

(A) are due to the leakage of small quantities of acetylcholine from a quiescent somatic efferent neuron
(B) are decreased in amplitude by anti-acetylcholine esterases (neostigmine, for example)
(C) are due to spontaneous depolarization of the motor end-plate in the absence of acetylcholine
(D) cause a series of weak muscle twitches
(E) cause a series of weak muscle tetanic contractions

59. Gamma-amino butyric acid (GABA) is believed to be one of the inhibitory transmitters in the spinal cord and brain. A likely mechanism of action for GABA is to

(A) increase the permeability of a neuron to Cl^-
(B) increase the permeability of a neuron to Na^+
(C) increase the permeability of the cell to Ca^{++}
(D) decrease the permeability of the cell to Ca^{++}
(E) increase the production of acetylcholine esterase

60. Curare relaxes muscle by

(A) decreasing the amplitude of the skeletal muscle end-plate potential
(B) preventing propagation of an action potential in skeletal muscle
(C) preventing propagation of an action potential in skeletal muscle and cardiac muscle
(D) enhancing the action of acetylcholine esterase
(E) enhancing the action of catecholamines (epinephrine, etc.)

61. Which of the following is characteristic of the generator potential of the pacinian corpuscle?

(A) it exhibits an all-or-none relationship with its stimulus
(B) it is a propagated phenomenon
(C) like the action potential, it exhibits an overshoot
(D) the amplitude increases rapidly at first, but then progresses less rapidly at high stimulus strength
(E) it shows extremely slow adaptation in the continued presence of a stimulus

Answers and Explanations

43. **(E)** Since this membrane is permeable to only K^+, its resting membrane potential will equal the K^+ equilibrium potential (E_K). E_K is determined by the activity gradient for K^+ that exists between the extracellular and intracellular fluids. It is the transmembrane potential that would occur when the net flux of K^+ across the membrane equals 0 if the cell membrane were permeable only to one ion, K^+. We can estimate E_K by using the *Nernst equation* and assuming that the concentration gradient for K^+, $(K^+_o)/(K^+_i)$, is approximately equal to the activity gradient for K^+:

$$E_K = 61.5 \log \frac{(K^+_o)}{(K^+_i)} = -97\text{mV}$$

If the concentration gradient for K^+ were 0.1, it would mean that K^+ was 10 times more concentrated inside the cell (K^+_i) than outside (K^+_o) and that E_K would equal –61.5 mV, since the log of 0.1 is –1. Since E_K equals –97 mv, the concentration of K^+ inside the cell must be more than 10 times that outside the cell, and therefore statements A and B must be wrong. When the resting potential is –97 mV, there is a separation of charge across the membrane, but we should still consider the cytoplasm and extracellular fluid electrically neutral, since this small separation of charge in the immediate region of the membrane does not measurably change the anion or cation concentration. An increase in sodium conductance (ie, permeability) will decrease the transmembrane potential (ie, move it toward E_{Na}, which is approximately +66 mV).

44. **(A)** The change of the transmembrane potential from –90 mV to +40 mV is due to an increased Na^+ conductance.

45. **(C)** Na^+ and Cl^- are more concentrated in the extracellular environment than in the intracellular environment, and thus their chemical gradients are into the cell. The opposite is true of K^+. The inside of the cell membrane is negative in reference to the outside, and thus the electrical gradient for positively charged ions is into the cell and the gradient for negatively charged ions is out of the cell.

46. **(E)** The electrogenic pump and differential permeability of the plasma membrane are both responsible for the resting transmembrane potential. Consistent with the pump hypothesis is the generally accepted conclusion that an active transport system extrudes 3 Na^+ from the cell for every 2 K^+ carried into the cell. Consistent with the permeability hypothesis is the conclusion that the plasma membrane during its resting potential is more permeable to K^+ than to Na^+. For example, the permeability coefficient for a resting skeletal muscle plasma membrane has been calculated to be 10^{-4} cm/sec for Cl^-, 10^{-8} cm/sec for K^+, and 10^{-10} cm/sec for Na^+. If such a cell had a Na^+/K^+ pump that carried approximately equal quantities of Na^+ out of and K^+ into the cell, the cell would lose more positively charged particles through diffusion than it would gain. It would therefore tend to develop a negative transmembrane potential if there were no compensating movement of other charged particles. The electrogenic pump potential (due to more Na^+ extruded than K^+ brought in) is only a small part of the resting potential.

47. (A) Donnan and Gibbs demonstrated that in the presence of nondiffusible ions, the diffusible ions distribute themselves so that the product of their concentrations on one side of the membrane is equal to the product of their concentrations on the other side of the membrane:

$$(K^+_i) \times (Cl^-_i) = (K^+_o) \times (Cl^+_o)$$
$$(120 \text{ mM}) \times (Cl^-_i) = (5 \text{ mM}) \times (120 \text{ mM})$$
$$(Cl^-_i) = 5 \text{ mM}$$

48. (D) Osmotic balance requires the same molar concentration of particles on both sides of the cell membrane. Therefore:

$$Na^+_i + K^+_i + Cl^-_i + \text{Protein}^-_i = Na^+_o + K^+_o + Cl^-_o$$
$$10 \text{ mM} + 120 \text{ mM} + 5 \text{ mM} + Pr^-_i = 115 \text{ mM} + 5 \text{ mM} + 120 \text{ mM}$$
$$Pr^-_i = 240 \text{ mM} - 130 \text{ mM} = 105 \text{ mM}$$

49. (D) In skeletal muscle cells the equilibrium potential for K^+ is approximately –97 mV when the transmembrane potential is –90 mV. Therefore, an increased permeability for K^+ would result in a K^+ efflux and a movement of the transmembrane potential toward –97 mV. This, by convention, is called an increase in the transmembrane potential.

50. (A) Record B has a reduced overshoot (30 mV vs. 60 mV). This could be due to a reduced concentration of Na^+ in the extracellular environment. Low K^+ would cause a more negative resting potential. High Na^+ would cause an increased overshoot. High K^+ would cause a less negative resting potential.

51. (E) Calcium is a structural part of the cell membrane and tends to stabilize it and therefore decrease its excitability. Impaired blood clotting occurs at Ca^{2+} levels below those causing lethal tetany.

52. (E) Smooth muscle, unlike other muscles, produces action potentials that do not necessarily have an overshoot. These potentials are mainly due to the influx of Ca^{++} and therefore are decreased in amplitude when the extracellular concentration of Ca^{++} is decreased. In order to answer this question correctly you must recognize that the slow wave potential has not changed, but the spike potentials have decreased in amplitude. In these records the slow waves on which the action potentials are superimposed are characteristic of many smooth muscles. Like the resting potential of skeletal muscle, they increase (ie, become more negative) by decreases in extracellular K^+ and decrease by increases in extracellular K^+. Apparently slow waves serve as a pacemaker potential. Note that as the slow wave becomes less negative it eventually initiates the development of spike potentials.

53. (C) The event responsible for the rapid repolarization of cells as different as neurons and cardiac cells, and always found during the repolarization of these cells, is a K^+ efflux (ie, movement of K^+ from the cell into the extracellular space). Most of the Na^+ gained by the cell during depolarization is moved from the cell by the Na^+ pump after repolarization.

54. (B) The influx of Na^+ during this period is due to an increased Na^+ conductance. Diffusion is a movement of particles along a concentration gradient. It is not energy requiring. Pinocytosis is a process of small vesicle formation whereby a small part of the external environment is brought into the cell. It is apparently important in bringing certain large molecules into the cell but plays little or no role in Na^+ transport. Interval h–i is caused by a diffusion of K^+ out of the cell. Most of the active transport of Na^+ and K^+ occurs after period h–i.

55. (D) The depolarization of the end-plate differs from the action potential found elsewhere in the muscle membrane in that the former is due to increased Na^+ and K^+ conductance and the latter is due to increased Na^+ conductance alone. Under normal conditions, the end-plate potential will change from –90 mV to –10 mV. This is not a reversal of charge but a decrease in the transmembrane potential. The end-plate potential, unlike the action potential, is a local (ie, nonpropagated) event.

56. **(D)** Under physiologic conditions, the end-plate potential is initiated by acetylcholine. In contrast, various excitatory neurotransmitters including acetylcholine can cause EPSPs. Both EPSPs and the end-plate potentials are nonpropagated depolarizations caused by an increased permeability of a part of the cell (end-plate, dendrites, cell body) to Na^+ and K^+.

57. **(E)** An increased g_{Cl^-} causes a diffusion of Cl^- to the cytoplasmic side of the cell membrane, resulting in membrane hyperpolarization. IPSPs occur in the CNS, spinal motor neurons, and other places.

58. **(A)** MEPPs are due to spontaneous leakage of acetylcholine resulting in subthreshold end-plate depolarizations. They do not cause muscle contractions.

59. **(A)** GABA causes IPSPs by increasing the permeability of a neuron to Cl^- resulting in membrane hyperpolarization and decreased excitability.

60. **(A)** Curare blocks the action of acetylcholine (ACh) at the neuromuscular junction. If sufficient quantities of curare are given, the amplitude of the end-plate potential becomes so small that it is incapable of initiating an action potential. The rest of the muscle is still excitable and will produce propagated action potentials in response to electrical stimulation.

61. **(D)** The generator potential, or receptor potential, like the excitatory postsynaptic potential, and the end-plate potential, (1) is a graded response (ie, not all-or-none), which (2) can initiate a propagated potential but is not itself propagated. (3) It will not exhibit an overshoot. The overshoot occurs when there is a markedly increased permeability to Na^+. Overshoot is characteristic of action potentials of axons and striated muscle cells. The generator potential of the pacinian corpuscle adapts extremely rapidly.

CHAPTER 4

Autonomic Nervous System

Questions

DIRECTIONS (Questions 62 through 68): Each set of matching questions in this section consists of a list of lettered options followed by a set of numbered words or phrases. For each numbered word or phrase, select the ONE lettered option that is most closely associated with it. Each lettered option may be selected once, more than once, or not at all.

Questions 62 through 65

(A) midbrain
(B) pons
(C) medulla oblongata
(D) cervical spinal cord
(E) thoracic spinal cord
(F) lumbar spinal cord
(G) sacral spinal cord
(H) preganglionic parasympathetic neuron
(I) postganglionic parasympathetic neurons
(J) preganglionic sympathetic neuron
(K) postganglionic sympathetic neurons
(L) adrenal medulla

62. No autonomic neurons originate from this part of the spinal cord.

63. These autonomic neurons originate from the brain stem.

64. These autonomic neurons originate from the thoracic and lumbar spinal cord.

65. This structure is innervated by preganglionic sympathetic neurons.

Questions 66 through 68

Match the following list of receptors with their agonists.

(A) nicotinic
(B) muscarinic
(C) α_1
(D) α_2
(E) β_1
(F) β_2

66. What type of receptor does epinephrine stimulate in the veins to produce venous constriction?

67. What type of receptor does epinephrine stimulate in the arterioles to produce arteriolar constriction?

68. What type of receptor does epinephrine stimulate in the arterioles to decrease arteriolar constriction?

DIRECTIONS (Questions 69 through 81): Each of the numbered items or incomplete statements in this section is followed by answers or by completions of the statement. Select the ONE lettered answer or completion that is BEST in each case.

69. Stimulation of postganglionic parasympathetic neurons causes all of the following EXCEPT

(A) release of acetylcholine
(B) an increased conduction time in the AV node of the heart
(C) relaxation of the internal anal sphincter
(D) contraction of the circular muscle of the iris
(E) secretion by the sweat glands

70. Stimulation of postganglionic sympathetic neurons causes all of the following EXCEPT

(A) an increase in heart rate and stroke volume
(B) vasodilation in the gastrocnemius muscle
(C) erection of the penis
(D) ejaculation in the male
(E) renin secretion

71. During an operation, a surgeon cuts all the neurons leading into the right and left sympathetic chain of ganglia. What are the immediate consequences of this operation?

(A) an inability of the arterioles of the arm to constrict
(B) an inability of the arterioles of the arm to constrict in response to a decrease in pressure in the carotid sinus
(C) an inability of the heart to speed up in response to a decrease in pressure in the carotid sinus
(D) an inability of the arterioles to constrict or the heart to speed up in response to a decrease in pressure in the carotid sinus
(E) a disappearance of intestinal peristalsis

72. Bilateral transection of the vagus nerves in the lower half of the neck causes all of the following EXCEPT

(A) a decreased ability to release renin in response to decreased right atrial pressure due to the section of sensory neurons
(B) a tendency to aspirate food and other material due to section of motor neurons to the extrinsic muscles of the larynx
(C) increased heart rate
(D) loss of the cephalic phase of gastric secretion
(E) loss of intestinal peristalsis due to the section of preganglionic parasympathetic neurons

73. Parasympathetic and sympathetic neurons exert approximately equal but opposite direct actions on

(A) the arterioles of the arm
(B) the force of ventricular contraction
(C) the frequency of ventricular contraction
(D) the pancreas
(E) sweat glands

74. Which of the following usually has norepinephrine as its major secretion?

(A) the adrenal medulla
(B) the adrenal cortex
(C) postganglionic sympathetic neurons to the radial muscles of the iris
(D) postganglionic sympathetic neurons to sweat glands
(E) postganglionic parasympathetic neurons to circular muscles of the iris

75. Atropine is a drug that blocks the action of acetylcholine on smooth muscles, glands, and the heart. Which of the following actions would you expect atropine to have?

(A) loss of control over the diaphragm
(B) cardiac asystole
(C) pupillary constriction
(D) decreased bronchial secretions
(E) failure of skeletal muscle vasoconstriction in response to the stimulation of sympathetic efferent neurons

76. Hexamethonium is a drug that blocks the action of acetylcholine on postganglionic autonomic neurons. Which of the following actions would you expect hexamethonium to have in a resting subject?

(A) a decrease in the heart rate
(B) an increase in the heart rate
(C) disappearance of intestinal peristalsis

(D) an increase in the arterial blood pressure
(E) prevention of skeletal muscle vasodilation in response to local hypoxia

77. Drugs that block the vasoconstrictor action of epinephrine in skeletal muscle are called α blockers. Drugs that block the vasodilator action of epinephrine in skeletal muscle are called β blockers. Norepinephrine has which of the following actions?

(A) it has the same action on all skeletal muscle arterioles as epinephrine
(B) it produces a less profound stimulation of the β receptors of arterioles than epinephrine and therefore produces a less profound increase in arterial pressure
(C) it produces a less profound stimulation of the β receptors of arterioles than epinephrine and therefore produces a more profound increase in arterial pressure
(D) it produces a less profound stimulation of the α receptors of arterioles than epinephrine and therefore produces a less profound increase in arterial pressure
(E) it produces a less profound stimulation of the α receptors of arterioles than epinephrine and therefore produces a more profound increase in arterial pressure

78. A patient who has been receiving phenoxybenzamine (an α-blocking agent) has complained of dizziness and fainting while walking up a single flight of stairs. During this exercise, his

(A) cardiac output would be similar to that before the exercise
(B) heart rate would increase
(C) systemic resistance would increase
(D) systemic resistance would fall to a lesser extent than before the drug was given
(E) venous tone would increase to the same extent as before the drug was given

79. Propranolol is a β-adrenergic blocking agent. It prevents the vasodilator and cardioaccelerator actions of epinephrine but does not prevent the vasoconstrictor action of either epinephrine or norepinephrine (see Fig. 4–1). These data are consistent with all of the following EXCEPT

(A) epinephrine produces vasodilation by stimulating β receptors
(B) epinephrine produces vasodilation by stimulating α receptors
(C) epinephrine produces vasoconstriction by stimulating α receptors
(D) epinephrine produces cardioacceleration by stimulating β receptors
(E) norepinephrine produces vasoconstriction by stimulating α receptors

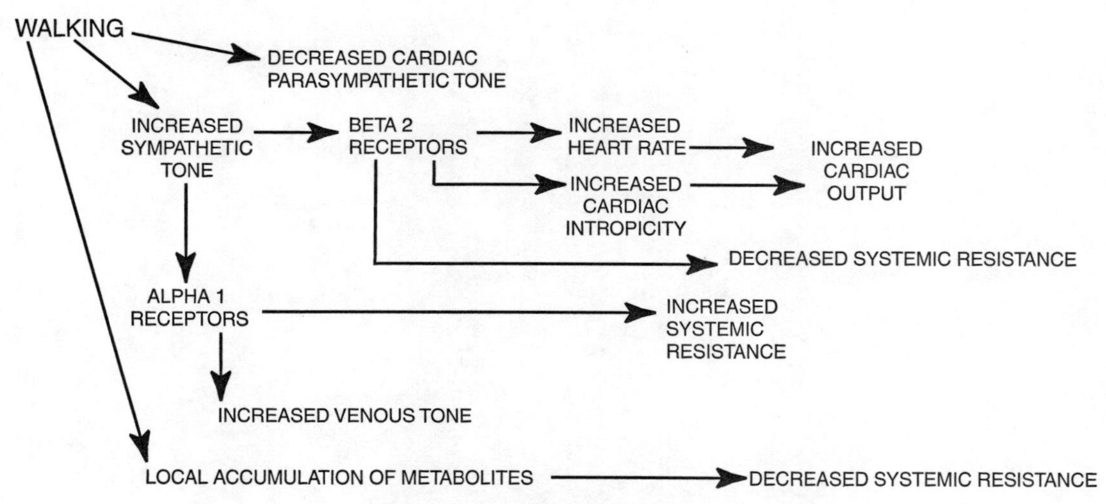

Figure 4–1. Effects of an alpha-blocking agent.

80. Norepinephrine and epinephrine both increase the heart rate and produce vasoconstriction in the skin. On the other hand, epinephrine produces vasodilation in skeletal muscle where norepinephrine produces vasoconstriction. These observations, as well as those presented in previous questions, are consistent with the view that

(A) norepinephrine stimulates only α receptors
(B) norepinephrine stimulates only β receptors
(C) norepinephrine stimulates α receptors of the heart and β receptors of the arterioles
(D) norepinephrine stimulates α receptors of the arterioles and β receptors of the heart, but not β receptors of the arterioles
(E) epinephrine stimulates only β receptors

81. Which of the following statements is correct?

(A) epinephrine produces a more marked increase in arterial pressure than norepinephrine
(B) norepinephrine produces a more marked increase in arterial pressure than epinephrine
(C) norepinephrine has the same β_2 stimulator action as epinephrine
(D) epinephrine produces a more extensive vasoconstriction than norepinephrine
(E) arterioles that have both α and β receptors will dilate in response to norepinephrine

Answers and Explanations

62–65. (62-D, 63-H, 64-J, 65-K and L)

66–68. (66-C, 67-C, 68-F) Epinephrine can cause either vasoconstriction or vasodilation, depending on the receptor type present on the blood vessel. Veins have predominantly α_1 receptors, while arteries have a mixture of α_1 and β_2 receptors.

69. (E) Sweat glands are innervated by postganglionic sympathetic neurons using acetylcholine as neurotransmitter.

70. (C) Erection of the penis is caused by activation of parasympathetic centers in the sacral spinal cord.

71. (B) There is no parasympathetic innervation of blood vessels (except for the penile artery). The only efferent neurons available for constricting and dilating their arterioles are the sympathetic neurons. Since all sympathetic neurons pass into the sympathetic chain, bilateral sympathectomy eliminates the reflex control of the blood vessels of the appendages. However, arterioles may still constrict in response to stretch and changes in their environment in the absence of a nervous system. One can cause a speeding up of the heart by either a stimulation of cardiac sympathetic neurons or an inhibition of cardiac parasympathetic neurons in the vagus nerve. The cranial parasympathetic neurons are left intact after this operation. Peristalsis is due to pacemakers in the intestine which continue to function in the absence of any connections with the central nervous system.

72. (E) The parasympathetic fibers in the neck originate from the brain stem and synapse with postganglionic neurons in the heart, alimentary tract, etc. The cervical vagi also carry sensory neurons from the cardiopulmonary receptors (important in the regulation of blood volume and sympathetic tone to the cardiovascular system), aortic receptors, lungs, etc. to the brain. Some of the motor neurons that pass down the vagi form the recurrent laryngeal nerves. They carry impulses up the neck to the striated muscles of the larynx. Intrinsic pacemakers in the intestine continue to cause peristalsis even after the destruction of the preganglionic fibers which innervate them.

73. (C) Parasympathetic neurons do not innervate structures in the arm, forearm, thigh, or leg. Arteriolar dilation in these appendages can be brought about by the inhibition of adrenergic sympathetic neurons. Parasympathetic neurons, unlike sympathetic neurons, exert little direct influence on the ventricles. They do play a major role in the control of the sinoatrial node, atrioventricular node, and atrial myocardium. The pancreas receives an important parasympathetic innervation but little innervation from sympathetic neurons.

74. **(C)** Sympathetic postganglionic neurons generally secrete norepinephrine. The adrenal medulla, unlike other producers of catecholamines in the body, has the capacity to methylate most of the norepinephrine it produces and, in this manner, convert it to epinephrine. The adrenal medulla usually has a secretion of about 80% epinephrine and 20% norepinephrine. During hypoxia or asphyxia, however, norepinephrine can constitute over 50% of the secretion. The adrenal cortex is an important producer of steroids (mineralocorticoids, glucocorticoids, androgens, and estrogens), but not of catecholamines. Human sweat glands and some arterioles are innervated by postganglionic, cholinergic, sympathetic neurons. All postganglionic parasympathetic neurons release acetylcholine, not catecholamines.

75. **(D)** Atropine-like agents are used prior to a general anesthesia to dry the mucous membranes of the respiratory tract. In preventing the accumulation of mucus in the tract, they reduce the occurrence of atelectasis. Atropine does not interfere with the stimulation of the diaphragm or other skeletal muscles. Acetylcholine modifies the intrinsic activity of the heart. Since it does not initiate contractions of the heart, blocking its action will not stop the heart from contracting. Acetylcholine causes pupillary constriction. Blocking its action would cause pupillary dilation. All postganglionic vasoconstrictor neurons are sympathetic fibers that release norepinephrine. Atropine does not block the action of norepinephrine, nor does it block the action of acetylcholine on postganglionic neurons.

76. **(B)** Although the heart rate of a resting subject may be controlled by a combination of sympathetic tone, which tends to speed the heart, and parasympathetic tone, which tends to slow the heart, the overall effect of the autonomic nervous system on heart rate in the resting subject is one of slowing it down. Therefore, when hexamethonium blocks the influence of the central nervous system on both sympathetic and parasympathetic neurons, the heart rate increases. Peristalsis requires functioning groups of intrinsic pacemakers but does not require signals from the central nervous system. The arterial blood pressure (P_a) is approximately equal to the cardiac output (I) times the peripheral resistance (Ω): P_a (mm Hg) = (I mL/min) × (Ω mm Hg/mL/min). Loss of autonomic tone causes some minor changes in cardiac output and marked arteriolar vasodilation (decreased peripheral resistance). In other words, the loss of adrenergic sympathetic tone to the arterioles causes a decreased arterial pressure. Hypoxia can produce arteriolar vasodilation in the absence of a nervous innervation.

77. **(C)** Epinephrine acts on both α and β receptors, while norepinephrine is more potent on α receptors. Therefore, norepinephrine produces a general vasoconstriction in skeletal muscle. Under most conditions, epinephrine produces vasodilation in skeletal muscle. Since norepinephrine is a general vasoconstrictor, it produces a more profound increase in arterial pressure than epinephrine.

78. **(B)** Under these conditions the cardiac output and heart rate will increase (β receptors in the heart are intact), and the decrease in systemic resistance will be exaggerated, because the patient has lost the constriction of the renal, visceral, and cutaneous arterioles that usually occurs during exercise, and the increased venous tone during exercise will be lost also. The dizziness and fainting during the exercise is apparently due to the low systemic resistance which causes an arterial hypotension and brain ischemia.

79. **(B)** Epinephrine produces vasoconstriction by stimulating α receptors and vasodilation by stimulating β receptors. Propranolol inhibits the action of epinephrine on β receptors.

80. **(D)** There are at least two types of β receptors. β_1 Receptors are found in the heart, liver, and adipose tissue, the β_2 receptors are found in smooth muscle. Norepinephrine is more potent on β_1 receptors than β_2 receptors.

81. **(B)** Norepinephrine acts mainly on α receptors, while epinephrine acts on both α and β receptors. Therefore, norepinephrine will cause a more marked increase in systemic resistance and arterial pressure than epinephrine.

PART III

Contraction

CHAPTER 5

Skeletal Muscle

Questions

DIRECTIONS (Questions 82 through 110): Each of the numbered items or incomplete statements in this section is followed by answers or by completions of the statement. Select the ONE lettered answer or completion that is BEST in each case.

82. Skeletal muscle contraction

(A) lasts approximately as long as the action potential for the muscle
(B) lasts approximately as long as the refractory period for the muscle
(C) A and B
(D) precedes the refractory period for the muscle
(E) none of the above

83. Skeletal and cardiac muscle are both striated and at resting length contain in each sarcomere an A band. This A band contains

(A) essentially all the contractile protein myosin, but no actin
(B) essentially all the contractile protein actin, but no myosin
(C) essentially all the myosin plus some actin
(D) essentially all the actin plus some myosin
(E) troponin and tropomyosin, but no actin

84. When skeletal muscle shortens in response to stimulation, there is

(A) a decrease in the width of the I band
(B) a decrease in the width of the A band
(C) a decrease in the width of the A and I bands
(D) an increase in the width of the H zone
(E) all of the above

85. During an isometric contraction, the length at which a single skeletal muscle sarcomere (Fig. 5–1) can exert its maximum force in response to stimulation is

(A) 1.7 μm
(B) 2.2 μm
(C) 3.0 μm
(D) all of the above
(E) both 2.2 and 3.0 μm

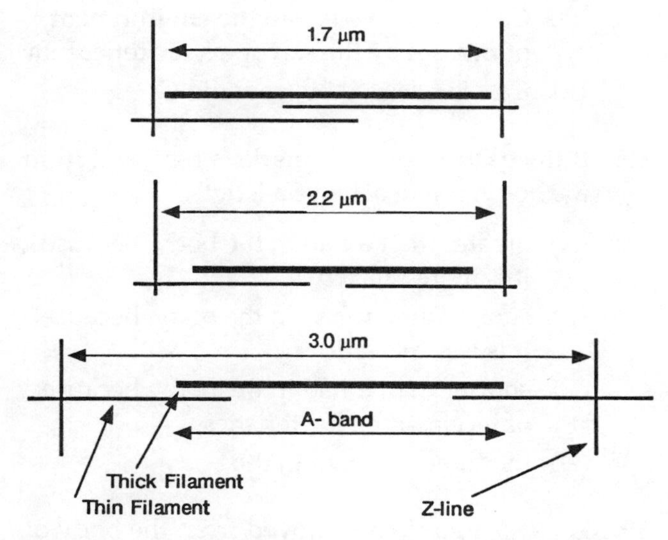

Figure 5–1. A single skeletal muscle sarcomere can exist at a number of lengths during the resting state.

86. Read each of the following statements about skeletal muscle contraction.

1. the major function of the transverse tubular system is to store and release Ca^{++}
2. the intracellular release of Ca^{++} causes the formation of bonds between actin and myosin
3. the bonds between actin and myosin are maintained until the Ca^{++} is sequestered

Which of the following BEST summarizes your conclusions?

(A) all of the above are true
(B) none of the above are true
(C) statement 2 is true
(D) statement 3 is true
(E) statements 2 and 3 are true

87. The action potential in striated muscle initiates a series of events that leads to a reaction between actin and myosin and the hydrolysis of adenosine triphosphate. This series of events is called the excitation–contraction coupling process. This is a process in which

(A) Ca^{++} is released into the immediate environment of the myosin by the sarcoplasmic reticulum
(B) Ca^{++} combines with troponin C
(C) both of the above occur
(D) Ca^{++} is removed from the environment of myosin by the sarcoplasmic reticulum
(E) Ca^{++} is released by troponin C

88. If the gastrocnemius muscle is removed from the body, it will achieve a length

(A) greater than it had in the body, because it is more relaxed
(B) shorter than it had in the body, because it is less relaxed
(C) shorter than it had ˙n the body, because of its elastic characteristics
(D) the same as it had in the body

89. A psoas muscle is removed from the body of a 70-kg man, allowed to obtain equilibrium length, stimulated with a maximal stimulus, and then the procedure of stretching and stimulation is repeated many times. If the muscle were stretched 2 additional millimeters each time and kept moist with isotonic saline, which of the following results would be expected? The muscle would

(A) not contract in response to stimulation when outside the body
(B) not contract in response to the 20th stimulus, because by then it would have been depleted of nutrients
(C) exhibit progressively more forceful contractions in response to stimulation as it is stretched the first five times
(D) exhibit progressively less forceful contractions in response to stimulation as it is stretched the first five times
(E) exhibit no change in force in response to stimulation since it is being stimulated by a maximal stimulus

90. Skeletal muscle has been characterized physiologically as having (a) a parallel elastic element (PE), (b) a series elastic element (SE), and (c) a contractile element (CE). Identify these in Figure 5–2A.

(A) 1, 2, 3
(B) 2, 1, 3
(C) 3, 2, 1
(D) 3, 1, 2
(E) 2, 3, 1

91. Muscle contractions may be isometric, isotonic, or associated with lengthening. Characterize the contractions in the muscle parts mentioned in Question 90 as seen in Figure 5–2B and C.

(A) isometric and isotonic
(B) isotonic and isometric
(C) isometric and lengthening
(D) isotonic and lengthening
(E) lengthening and isotonic

92. Which of the following statements is correct?

(A) a muscle at resting length exerts its maximum force during an isotonic contraction

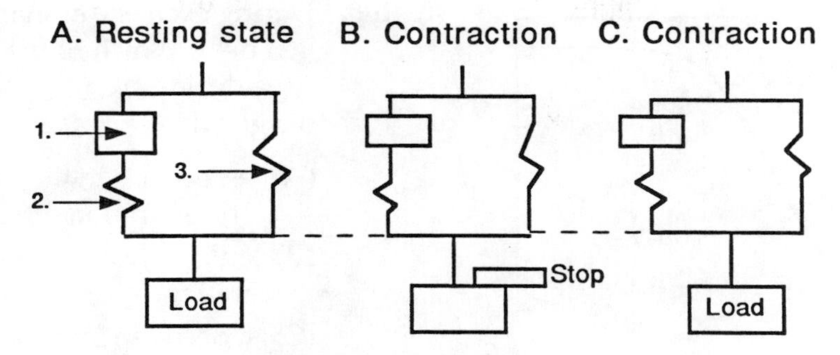

Figure 5–2. Schematic diagram of muscle.

(B) the maximum velocity of shortening during contraction occurs when there is no afterload
(C) the preload is the weight the muscle moves before it starts to relax
(D) in most forms of muscle contraction in an intact individual, the preload and afterload are equal
(E) none are correct

93. Which of the following best defines contraction?

(A) a series of chemical reactions that cause the muscle to pull
(B) a series of chemical reactions that cause the muscle to shorten
(C) a series of chemical reactions in which the muscle responds to stimulation
(D) shortening
(E) production of tension

94. An isotonic contraction differs from an isometric contraction in that in an isotonic contraction

(A) the muscle is less efficient
(B) the muscle uses more high-energy phosphate bonds
(C) the heat of activation is greater
(D) the recovery heat is reduced
(E) the heat of activation is less

95. A skeletal muscle participating in a strenuous exercise differs from a resting skeletal muscle in that the exercising muscle

(A) releases markedly larger quantities of lactic acid into the blood that passes through it
(B) transmits markedly increased quantities of heat to the blood passing through it
(C) exhibits both of the above characteristics
(D) has a venous blood with a markedly reduced concentration of O_2
(E) exhibits all of the above characteristics

96. The rate at which Ca^{++} is sequestered by the sarcoplasmic reticulum of skeletal muscle during a twitch is directly related to

(A) the rate of tension development
(B) the rate of ATP hydrolysis by myosin
(C) both of the above
(D) the height of the action potential
(E) the rate of relaxation

97. In a series of experiments, it is noted that in a skeletal muscle fiber an intracellular concentration of Ca^{++} of $10^{-6.5}$ mol/L is the threshold value needed for inducing contraction (see Fig. 5–3 on page 40). On this basis, one would expect a concentration of $10^{-5.5}$ mol of Ca^{++} per L to cause

(A) a more forceful contraction
(B) a less forceful contraction
(C) a contraction of equal force
(D) relaxation

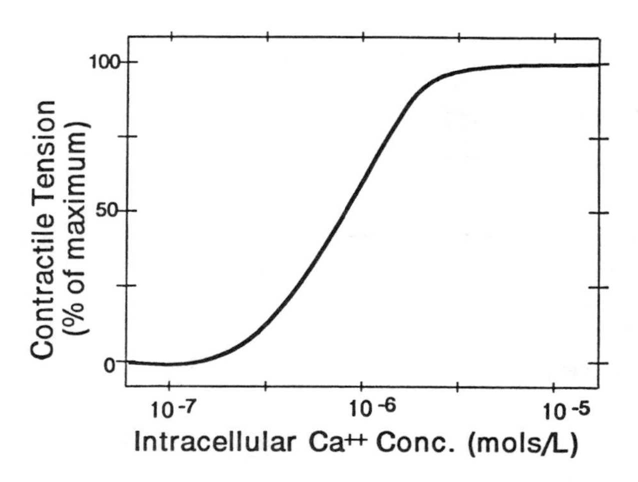

Figure 5–3. Contractile tension as a function of intracellular Ca^{++} concentration.

98. The all-or-none law stated by Bowditch for the striated muscle cell is that the strength of contraction of a single cell responding to a single stimulus

(A) cannot be increased in a healthy subject
(B) cannot be changed by changing the environment of the cell
(C) cannot be increased by increasing the strength of the stimulus (usually an electrical stimulus) above threshold
(D) cannot be increased by increasing the frequency of stimulation
(E) is not increased in hypertrophy

99. Which of the following statements best defines the treppe phenomenon described by Bowditch?

(A) a maintained contraction associated with multiple action potentials
(B) an increase in the force generated in a twitch due to an increased frequency of stimulation
(C) a decrease in the force generated by a muscle due to a prolonged period of stimulation
(D) a failure to relax following stimulation
(E) a maintained bond between the myosin cross-bridges and actin that is not associated with an action potential

100. During exercise, human beings create an oxygen debt. Which of the following changes occurs during exercise and contributes to the oxygen debt?

(A) a decrease in the arteriovenous oxygen concentration difference
(B) an increase in the arterial lactate concentration
(C) a decrease in the arterial O_2 concentration
(D) a decrease in the concentration of hemoglobin
(E) an increased concentration of CO_2 in the venous blood

101. Which of the following statements concerning adult skeletal muscle is correct? Skeletal muscle responds to decentralization

(A) with an increase in tone
(B) by a progressive atrophy that eventually leads to the muscle's inability to contract in response to stimulation
(C) by muscle fiber hyperplasia on reinnervation
(D) with an increased sensitivity to acetylcholine during the first 2 months of decentralization
(E) B and D are correct

102. Which of the following characteristics of skeletal muscle make tetanic contraction possible?

(A) the motor neurons to skeletal muscle have a short refractory period and are therefore capable of delivering a high frequency of stimuli to a muscle fiber
(B) the cell membrane of the skeletal muscle fiber recovers its excitability well before the cell ceases its contraction
(C) both A and B are correct
(D) the prolonged exposure of the muscle end plate to high concentrations of acetylcholine throughout the tetanus
(E) the action potential of skeletal muscle outlasts the period of contraction

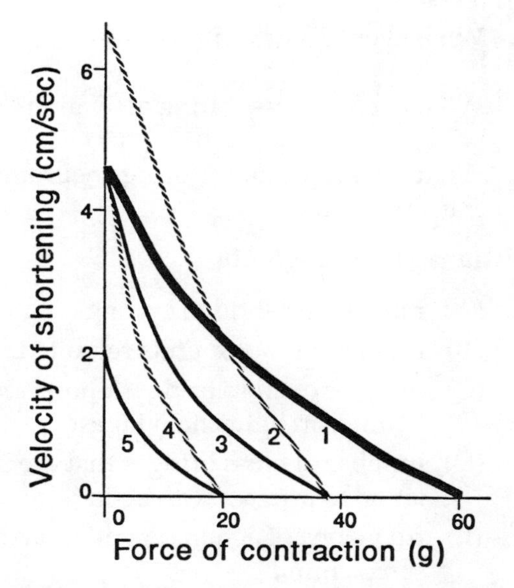

Figure 5–4. Force–velocity curves for skeletal muscle.

103. The force–velocity curves in Figure 5–4 were obtained from a muscle. Assume that supermaximal stimuli were used in each study, and assume that curve 1 was obtained for a muscle at L_{max} (ie, the muscle length at which the greatest active tension is developed) and that the other curves were obtained at a muscle length less than L_{max}. What is V_{max} for curve L?

(A) greater than 60 g
(B) less than 60 g
(C) greater than 4 cm/sec
(D) less than 4 cm/sec
(E) less than 2 cm/sec

104. Which curve in Question 103 would be caused by the sarcoplasmic reticulum releasing less Ca^{++} in response to stimulation of the muscle?

(A) curve 1
(B) curve 2
(C) curve 3
(D) curve 4
(E) curve 5

105. Which curve(s) in Question 103 represent(s) the muscle with the smallest preload?

(A) curve 1
(B) curve 3
(C) curve 4
(D) curves 2 and 3
(E) curves 2 and 5

106. The following length–tension diagram (Fig. 5–5) was obtained for a muscle. Supermaximal tetanic stimuli were used to initiate a contraction at each muscle length studied. Which point represents a preload of 40 g?

(A) point 3
(B) point 4
(C) point 8
(D) points 4 and 8
(E) points 3, 4, and 8

107. Maximal active tension in Figure 5–5 is developed by skeletal muscle at point(s)

(A) 1
(B) 2
(C) 4
(D) 3 and 4
(E) 9

108. Which point(s) in Figure 5–5 represent(s) no overlap between most of the muscle's thick and thin filaments?

(A) point 2
(B) point 3
(C) point 6
(D) point 7
(E) point 6 and 7

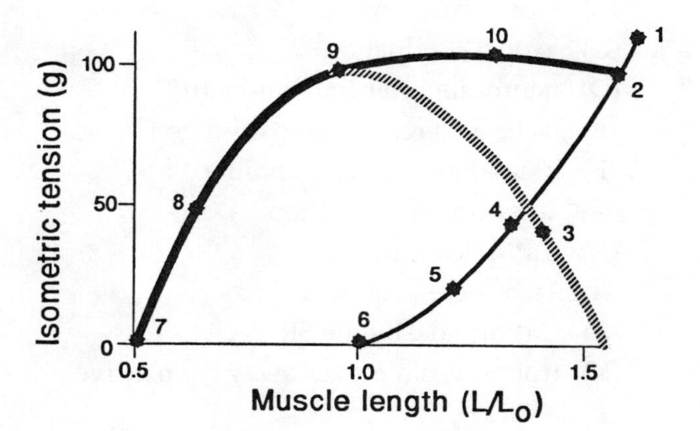

Figure 5–5. Length–tension curve for muscle.

109. Which of the following is true regarding the differences between the slow twitch muscle fiber and the fast twitch glycolytic fiber?

(A) smaller number of muscle fibers in each motor unit
(B) higher concentration of myoglobin and mitochondria
(C) higher ATPase activity
(D) A and B are correct
(E) A, B, and C are correct

110. Which of the following is correct regarding the differences between the slow twitch muscle fiber and the fast twitch glycolytic fiber?

(A) in large limb muscles serves as a reserve which can be recruited if there is a forceful contraction
(B) is more readily fatigued
(C) is part of a motor unit that consists of only red fibers
(D) A and C are correct
(E) A, B, and C are correct

DIRECTIONS (Questions 111 through 116): Each set of items in this section consists of a list of lettered options followed by several numbered words or phrases. For each numbered word or phrase, select the ONE lettered option that is most closely associated with it. Each lettered option may be selected once, more than once, or not at all.

Questions 111 through 113

(A) action potential in muscle membrane
(B) cross-bridge swivel
(C) axonal conduction
(D) neuromuscular transmission
(E) lactic acid reformed to glycogen
(F) breakdown of acetylcholine
(G) cross-bridge formation
(H) Ca^{++} release by SR
(I) sarcomere lengthening
(J) Ca^{++} uptake by the SR
(K) tropomyosin moves away from active site
(L) Ca^{++} combines with troponin C

111. What event occurs with recovery?

112. What event occurs during activation?

113. What event occurs during neuromuscular transmission?

Questions 114 through 116

(A) rate of cross-bridge cycling
(B) number of motor units recruited
(C) number of cross bridges phosphorylated by myosin light chain kinase
(D) number of cross bridges that are in a position to interact with actin
(E) frequency of action potentials across the sarcolemma
(F) rate at which the series elastic component is stretched
(G) rate of Ca influx across the sarcolemma
(H) rate of Ca release from the sarcoplasmic reticulum

114. During an isotonic contraction of smooth muscle, the maximum tension can vary with the type of stimulus and is a function of what?

115. During an isometric contraction of skeletal muscle, the maximum amount of tension that can be developed is determined by which of the above?

116. During an isotonic contraction of skeletal muscle, the maximum rate of shortening is a function of what?

Answers and Explanations

82. **(E)** The action potential characteristically lasts about 2 to 4 msec, and the period of contraction ranges between 5 and 50 msec. The muscle is refractory for most of the action potential.

83. **(C)** Much of the actin lies in the I band, and all the myosin is in the A band.

84. **(A)** During shortening, there is a greater overlap between actin and myosin and, as a result, a shorter I band. The A band retains approximately the same length. The H zone is formed by the ends of the actin filaments. Because these ends come closer together in shortening, the H zone becomes narrower.

85. **(B)** These diagrams are based on the sliding filament hypothesis. A sarcomere exerts its maximum contractile force at a length of 2.2 μm because apparently, at this length, there is an optimal relationship between the myosin cross-bridges on the thick filament and the actin receptor sites on the thin filament. At 1.7 μm, some cross-bridges may oppose the action of others, because the thin filament now extends to the contralateral side of the thick filament. At 3.0 μm, there are fewer actin receptor sites available to the myosin cross-bridges.

86. **(C)** In skeletal muscle, most of the calcium is stored and released by the sarcoplasmic reticulum. The T system is a continuation of the sarcolemma deep into the center of the cell and probably has as its major function the rapid transmission of the action potential from the cell surface. A maintained bond between actin and myosin causes rigor. In contraction, there is a continuous making and breaking of bonds. This is the process that ends when Ca^{++} is sequestered.

87. **(C)** An increase in the concentration of unbound Ca^{++} in the intracellular fluid, a combining of some of that Ca^{++} with troponin C, and a resultant shift of tropomyosin and tropin (ie, troponin I, C, and T) away from the actin receptor site are important parts of the excitation–contraction coupling process in striated and smooth muscle. In the case of skeletal muscle, most of the Ca^{++} is delivered by the terminal cisterns of the sarcoplasmic reticulum. In smooth muscle and the heart, the cell membrane and external environment are more important sources of Ca^{++} than in striated muscle.

88. **(C)** The shortening of muscle, in this case, is not associated with an action potential, an intracellular release of Ca^{++}, a heat of activation, a heat of shortening, or a chemical reaction between actin and myosin, and therefore is not a contraction in the sense this term is used in muscle physiology.

89. **(C)** Skeletal muscle maintains its excitability and capacity to contract outside the body as long as it still has nutrient reserves and is not destroyed by careless handling. Although it has a very limited reserve of O_2 in its myoglobin, it does have large reserves of glycogen which can be catabolized to lactic acid and energy in the absence of O_2. In the body, for example, skeletal muscle will remain excitable for up to 2 hours after its blood supply has been cut off. Organs like the cerebral cortex, which have little or no glycogen reserves, stop functioning after 20 seconds of ischemia. Within limits, stretching skeletal muscle increases its force of contraction.

90. (C) The contractile elements include actin and myosin. The series elastic elements lie in an end-to-end relationship with the CEs and may include the tendon and cross-bridges. The parallel elastic element lies in a side-by-side relationship with the CEs and includes the sarcolemma.

91. (A) In an isometric contraction, the muscle does not change its length. The contractile element in the model shortens, and the series elastic element is stretched. In an isotonic contraction, the whole muscle and the contractile element shorten.

92. (B) A muscle is capable of a greater contractile force during an isometric contraction. The more a muscle shortens during contraction, the less force it is capable of exerting. The greater the afterload, the slower the velocity of shortening. The afterload is equal to the load that opposes muscle shortening during contraction. The preload is the stretching force to which the muscle is exposed before it starts to contract. In most skeletal muscle contractions, the preload is close to 0 and the load lifted during contraction (the afterload) exceeds 0.

93. (A) Although contraction is frequently associated with shortening (isotonic contraction), it can also be associated with no change in length (isometric contraction) or even with an increase in length. This latter occurs when a muscle is contracting but its antagonists are contracting more forcefully. The elastic elements of muscle are capable of producing tension when they are stretched. This tension is called passive tension to distinguish it from active tension, the tension caused by contraction.

94. (B) Efficiency is defined as: (work performed in cal)/(total energy expenditure in cal). If you define work as: (force in dynes) (distance in cm) (2.39 cal/dyne^{-cm}), then the efficiency in an isometric contraction is 0, since there is no distance moved and therefore no work performed. The efficiency of an isotonic contraction may go as high as 40%. A muscle that shortens releases more heat than one that does not. This additional amount of heat is called the heat of shortening and is related to the amount of shortening in centimeters. This heat will come from the hydrolysis of high-energy phosphate bonds. The heats of activation for isometric and isotonic contractions are about equal. The recovery heat is elevated.

95. (E) During exercise, skeletal muscle produces more heat and lactate than during rest. These increases in production are reflected in the venous blood. Although the blood flow to skeletal muscle during exercise increases, this, in itself, is not enough to meet the increased needs of the muscle for O_2. Therefore, in response to exercise, the arteriovenous (AV) O_2 concentration difference may change from 5 mL/100 mL of blood to 15 mL/100 mL of blood. The heart differs from skeletal muscle in that, in the resting individual, it has an AV O_2 difference of 13 mL/100 mL (ie, one approximately equal to that for skeletal muscle during strenuous exercise).

96. (E) These phenomena are directly related to the concentration of Ca^{++}, not to its removal by the sarcoplasmic reticulum.

97. (A) Within limits, increases in the intracellular Ca^{++} levels above threshold will determine the number of troponin molecules combining with Ca^{++} per msec and hence will cause progressively more forceful contractions. Since increases in extracellular Ca^{++} cause increases in intracellular Ca^{++} in the sarcoplasmic reticulum, they can also cause increases in contractile force.

98. (C) This law applies to neurons as well as to muscle fibers. In the case of a single cell, you cannot change the characteristics of the action potential or the contraction by increasing the strength of the stimulus beyond threshold. Changes in the environment of the cell, increased frequencies of stimulation, and hypertrophy can all increase the force of contraction and/or the amplitude of the action potential.

99. **(B)** This is a definition of a complete tetanus. Albert Szent-Gyorgyi defined treppe as a condition in which activity creates a situation favorable for activity. This is sometimes referred to as a warm-up phenomenon and is presumably due to a change in the environment. When activity creates an environment unfavorable for activity, fatigue occurs. This is called contracture.

100. **(B)** During and after exercise, there is usually an elevated O_2 consumption. The fact that O_2 consumption immediately after exercise does not return to resting values is the evidence that human beings, during exercise, create an oxygen debt. During exercise, there is an increased removal of O_2 from the blood that causes an increased AV O_2 concentration difference. This occurs even though the blood flow through active skeletal muscle also increases. During exercise, skeletal muscle depends increasingly on the anaerobic catabolism of glucose to lactate and, as a result, sends large quantities of lactate into the blood. After exercise, the O_2 consumption of the body will remain elevated until the lactate in the blood is returned to normal levels. Part of the lactate will be aerobically catabolized to CO_2, water, and energy. Part of this energy will be used to convert some of the remaining lactate to glucose and glycogen. The healthy heart, unlike skeletal muscle, does not add lactate to the blood. In most exercise, the arterial concentration of O_2 stays fairly constant. Hemoglobin does not decrease in concentration in exercise. A decrease in the concentration of hemoglobin has little or no effect on O_2 consumption.

101. **(E)** Decentralization of skeletal muscle causes flaccidity. After death, there may also be rigor, but this is due to the disappearance of ATP, not the destruction of nerves. Eventually, all the contractile proteins in the muscle are catabolized. Reinnervation may cause hypertrophy of the remaining viable muscle fibers, but there is no evidence that it will cause hyperplasia (production of more cells). The stimulation of intact cholinergic autonomic neurons may cause twitches in decentralized skeletal muscle due to its hypersensitivity to acetylcholine. The reason for this may be that decentralized muscle contains less acetylcholine esterase than normal muscle.

102. **(C)** It is through the delivery of a high frequency of stimuli to an excitable cell that we produce a maintained contraction. This requires that the refractory period of both the motor neuron and the muscle fiber be markedly shorter than the period of contraction that occurs in a twitch. The acetylcholine released by a neuron is practically all destroyed or removed before the next action potential causes more to be released. In other words, the accumulation of acetylcholine at the neuromuscular junction is not the mechanism for the production of a maintained contraction. If the action potential outlasted the period of contraction, the refractory period would be too long to permit a fusion of twitch contractions.

103. **(C)** V_{max} is the maximum velocity of shortening for a muscle. It occurs when the afterload equals 0. In Figure 5–4, the afterload is represented by the force of contraction. V_{max} is determined by measuring the velocity of shortening for a muscle at different afterloads, plotting the data, and extrapolating the plot to 0 afterload.

104. **(E)** A decrease in available Ca^{++} is an example of a negative inotropic condition. Any negative inotropic agent will shift the force–velocity curve down and decrease V_{max}.

105. **(C)** At muscle lengths below L_{max}, the force of contraction during an isometric contraction will be directly related to the preload and inotropicity. The isometric force for each curve is determined by noting where each curve intersects the 0 velocity of shortening line. This is sometimes called P0 (pressure, tension, or force at which there is 0 shortening). In other words, curve 1 represents a muscle at an optimal preload, and curves 2, 3, and 4 represent muscles with a lower preload. Curve 5 (low V_{max}) differs from curve 4 (intermediate V_{max}) in that it represents a muscle that was exposed to a negative inotropic condition. Curve 2 (high V_{max}), on the other hand, is from a muscle exposed to a positive inotropic condition. If curve 5 represented the same inotropicity as curve 4, it would touch the 0 velocity line at a force greater than 20 g. Variations in preload have little effect on V_{max}. Variations in inotropicity have a marked effect on V_{max}.

106. **(B)** Line 6–1 represents the passive tension. At point 4, the passive tension is 40 g, and the muscle length is 1.3 times resting length. Passive tension and preload are synonymous.

107. **(E)** Active tension (line 7–9–3) is equal to total tension (line 7–9–2–1) minus passive tension (line 6–1).

108. **(A)** As you stretch a muscle from one-half its resting length (L0) toward twice its resting length, you reach a point (2) where the muscle no longer develops active tension in response to stimulation. This is the point at which the actin of the thin filament can no longer react with the myosin of the thick filament.

109. **(D)** Skeletal muscle fibers in humans can be divided into three categories: (I) slow oxidative, (IIA) fast oxidative, (IIB) fast glycolytic. The slow oxidative fiber will have a lower velocity of contraction, a less rapid relaxation, and a greater twitch duration than the type II fiber. Each motor unit will contain only one type of fiber. The motor unit with type I fibers innervates a smaller number of muscle fibers and exerts less force. In a weak contraction only type I units are activated. In a strong contraction both type I and II units are activated, with the latter exerting most of the force. The high ATPase activity of both type II fibers is, in part, responsible for their rapid fatigue during contraction. The type IIA fiber is the least common in the human. It is similar to the type I fiber and different from the type IIB in that it has a high capacity for oxidative catabolism, a high concentration of myoglobin and mitochondria, and a high capillary density.

110. **(C)** Same explanation as 109.

111–113. (111-E, 112-A, 113-F)

Neuromuscular transmisssion:

a. release diffusion, and destruction of acetylcholine
b. production of an end-plate potential

Activation:

a. production and propagation of an action potential in the plasma membrane
b. propagation of the action potential along the transverse tubules

Activation–contraction coupling:

a. release of Ca^{++} by sarcoplasmic reticulum
b. combination of Ca^{++} with troponin C
c. movement of troponin and tropomyosin away from actin receptor site

Contraction:

a. formation of cross-bridges between actin and myosin
b. cross-bridges swivel and produce shortening of sarcomere and/or a force

Relaxation:

a. sequestering of Ca^{++} in the sarcoplasmic reticulum
b. blocking of actin receptor site by troponin and tropomyosin
c. sarcomere lengthens and/or force of contraction decreases

Recovery:

a. creatine + phosphate → creatine phosphate
b. lactic acid + $O_2 \rightarrow CO_2 + H_2O$ + glycogen

114. **(G)**

115. **(B)**

116. **(A)**

CHAPTER 6

Cardiac Muscle

Questions

DIRECTIONS (Questions 117 through 152): Each of the numbered items or incomplete statements in this section is followed by answers or by completions of the statement. Select the ONE lettered answer or completion that is BEST in each case.

The following transmembrane potentials were recorded from healthy smooth muscle, cardiac muscle, and skeletal muscle (Fig. 6–1). They each include the resting transmembrane potential followed by an action potential. All were recorded at the same paper speed and sensitivity. Refer to this diagram for Questions 117, 118, and 119.

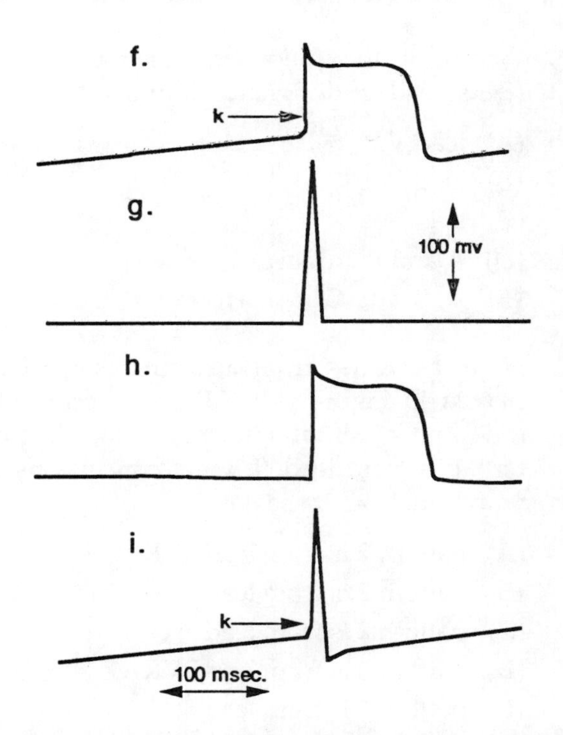

Figure 6–1. Resting membrane potentials from various kinds of muscle.

117. Which of the potentials in the diagram were recorded from potential pacemaker cells?

(A) cells f and h
(B) cells g and i
(C) cells f and i
(D) cells g and h
(E) cell f is the only potential pacemaker cell

118. Point k in patterns f and i in the diagram represents

(A) the threshold potential for the cell
(B) the end of the cell's refractory period
(C) the beginning of the period of supernormal irritability
(D) the end of the refractory period and the beginning of hyperirritability
(E) the cell's rheobase

119. Which of the potentials in the diagram might have been recorded from the heart?

(A) cells f and h
(B) cells g and i
(C) cells g and h
(D) cell h
(E) cells f, g, h, and i

120. Which of the following is not characteristic of the heart?

(A) its contraction is initiated by a nerve impulse
(B) it conducts impulses from one muscle cell to the next
(C) it contains a number of cells with an unstable transmembrane potential
(D) it contains a number of cells with a stable transmembrane potential
(E) its ventricles are inexcitable for most of their period of contraction

121. The sinoatrial node is the pacemaker for the heart because it

(A) is the most richly innervated structure in the heart
(B) is the only structure in the heart capable of generating action potentials
(C) has the highest rate of automatic discharge
(D) has the most stable transmembrane potential
(E) is the cardiac cell least sensitive to catecholamines

122. The most slowly conducting part of the heart is the

(A) atrium
(B) ventricle
(C) Purkinje fibers
(D) right bundle branch
(E) the atrioventricular junctional tissue at the atrioventricular node

123. The last part of the ventricles to be activated after atrial activation is

(A) the base of the right ventricle
(B) the base of the left ventricle
(C) the endocardium of the right ventricle
(D) the epicardium of the apex
(E) the endocardium of the apex

124. The heart has been characterized as a functional atrial syncytium connected by an atrioventricular conducting system to a functional ventricular syncytium. What would happen if a drug such as quinidine or procainamide were given that prevented intercellular conduction?

(A) there would be an increase in cardiac parasympathetic tone
(B) multiple ectopic pacemakers would develop
(C) both of the above would occur
(D) the force of ventricular contraction would increase
(E) all of the above would occur

125. In standard electrocardiogram lead III, the right leg electrode is grounded through the recorder and the potential difference is recorded between

(A) the right arm and left leg
(B) the right arm and left arm
(C) the left arm and left leg
(D) the right arm and a central terminal using resistors and the left arm and left leg
(E) the left arm and a central terminal using resistors and the right arm and left leg

126. Which of the following electrocardiogram (ECG) leads is designated unipolar?

(A) lead V2
(B) lead aVL
(C) lead I
(D) A and B are correct
(E) A, B, and C are correct

127. Note the ECGs from a normal subject displayed in Figure 6–2. All were recorded at the same sensitivity and paper speed. If pattern 1 is from lead II, what conclusions can you draw from these data?

(A) pattern 2 is from lead aVR
(B) pattern 2 is from lead aVF
(C) pattern 2 is from lead V4
(D) pattern 3 is from lead aVR
(E) pattern 3 is from lead I

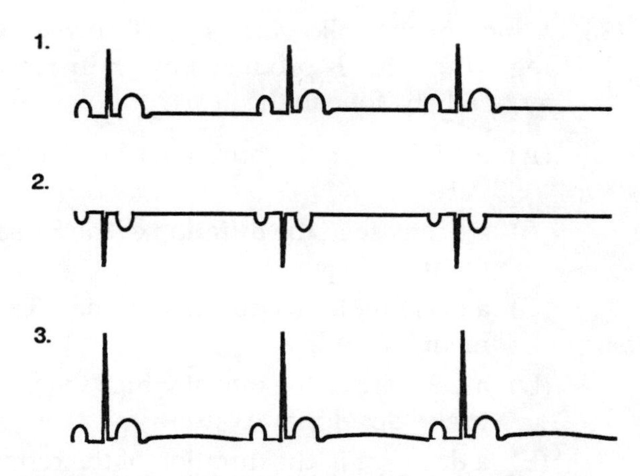

Figure 6–2. Electrocardiograms from a normal subject.

128. The depolarization of the atrium consists of numerous waves of depolarization (ie, waves of negativity) moving away from the sinoatrial node. These waves attract positive charges. It proves useful to characterize all the electrical changes in the heart at any one instant in terms of a single, two-dimensional dipole. This dipole has a positive head and a negative tail. The negativity represents the wave of depolarization. On that basis, one would expect the atrial dipole on the frontal plane to be moving

(A) cephalad (ie, –90°)
(B) caudad (ie, +90°)
(C) toward the right arm (180°)
(D) toward +60°
(E) toward –60°

129. If the frontal plane atrial dipole had an axis of +90°, the highest amplitude P wave on the ECG would be seen on lead

(A) I
(B) II
(C) III
(D) aVL
(E) aVF

130. The R wave of the ECG from a healthy subject represents a dipole with an electrical axis of

(A) –120°
(B) –60°
(C) 0°
(D) 60°
(E) 120°

131. A patient was admitted to the hospital and ECG lead III was recorded. It was found to contain no S wave. The P, R, and T waves appeared normal. What conclusions can you draw?

(A) activation of parts of the base of the heart are abnormal
(B) activation of parts of the apex of the heart are abnormal
(C) there has been cardiac depression
(D) there is left bundle branch block
(E) there are no indications of cardiac abnormalities

132. The following ECG (Fig. 6–3) was recorded from lead aVF. How would you characterize the pattern?

(A) a sinus rhythm
(B) a ventricular rhythm
(C) a nodal rhythm
(D) flutter
(E) fibrillation

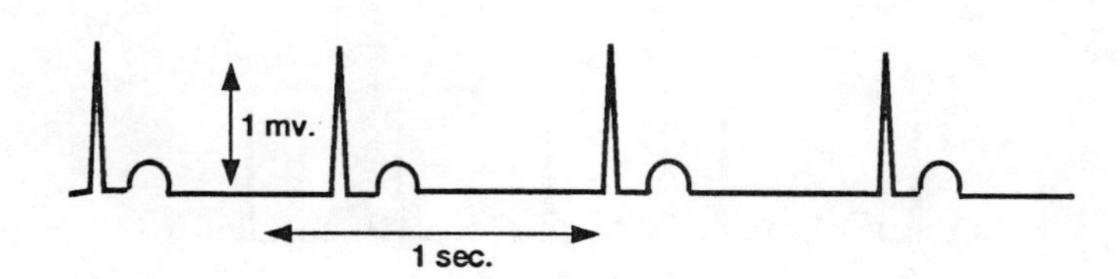

Figure 6–3. Electrocardiogram recorded from lead aVF.

133. A patient had an ECG with a prolonged PR segment. What might have caused this?

(A) increased sympathetic tone to the atrioventricular (AV) node
(B) increased parasympathetic tone to the AV node
(C) increased sympathetic tone to the sinoatrial (SA) node
(D) increased parasympathetic tone to the SA node
(E) decreased carotid sinus pressure

134. The T wave of the ECG occurs

(A) at the beginning of the heart's refractory period
(B) during the depolarization of the heart
(C) during atrial systole
(D) during the repolarization of the ventricle
(E) during the first heart sound

135. Which of the following events may either cause or be a sign of nodal rhythm?

(A) increased sympathetic tone to the SA node
(B) increased parasympathetic tone to the SA node
(C) a prolonged QRS complex
(D) a prolonged PR segment
(E) B and D

136. The following ECG (Fig. 6–4) was recorded from lead II of a patient. How would you characterize this pattern?

(A) complete heart block
(B) second-degree block
(C) sinus arrhythmia
(D) ventricular tachycardia
(E) ventricular bradycardia

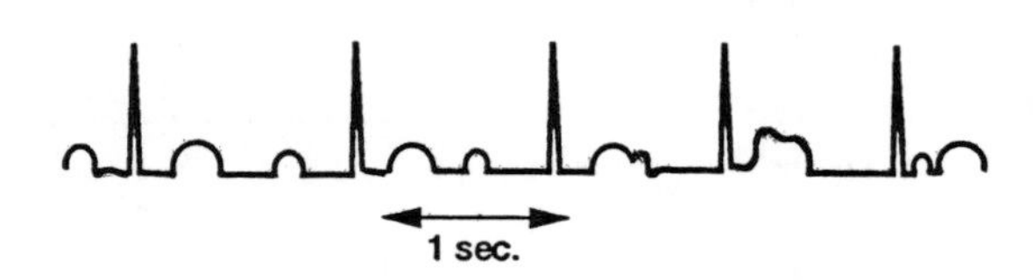

Figure 6–4. Electrocardiogram recorded from lead II.

137. Which of the following events does the pattern in Figure 6–4 suggest?

(A) the heart has two separate pacemakers
(B) ventricular conduction is abnormally slow
(C) there is atrial asystole
(D) A and B
(E) A and C

138. Which of the following is NOT correct with regard to what is produced by an increase in sympathetic tone to the heart?

(A) a decrease in the duration of R-R′ interval
(B) a decrease in the duration of the ST segment
(C) a decrease in the duration of the PR segment
(D) a decrease in the rate at which the ventricles develop pressure
(E) a decrease in the duration of the refractory period of the AV node

139. Consider what type of ECG would be caused by each of the other abnormalities listed above. An independence of the P wave and the QRS complex of the ECG indicates

(A) an early repolarization of the ventricular fibers
(B) a failure of the AV node to conduct
(C) depression of the SA node
(D) slowing of conduction at the AV node
(E) a conduction block in the left bundle branch

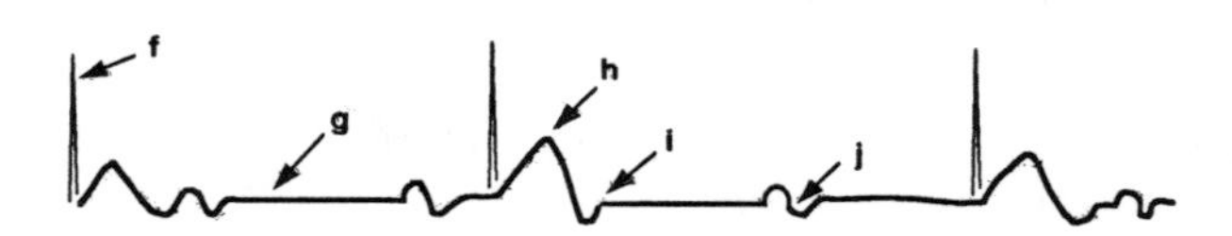

Figure 6–5. Electrocardiogram obtained in lead II from a patient with an electronic pacemaker.

140. Examining Figure 6–5, identify the following: (1) isoelectric line, (2) R wave, (3) T wave, (4) pacemaker artifact.

(A) 1: g, 2: f, 3: h, 4: j
(B) 1: j, 2: f, 3: i, 4: h
(C) 1: g, 2: h, 3: i, 4: f
(D) 1: g, 2: h, 3: f, 4: j
(E) 1: g, 2: f, 3: i, 4: h

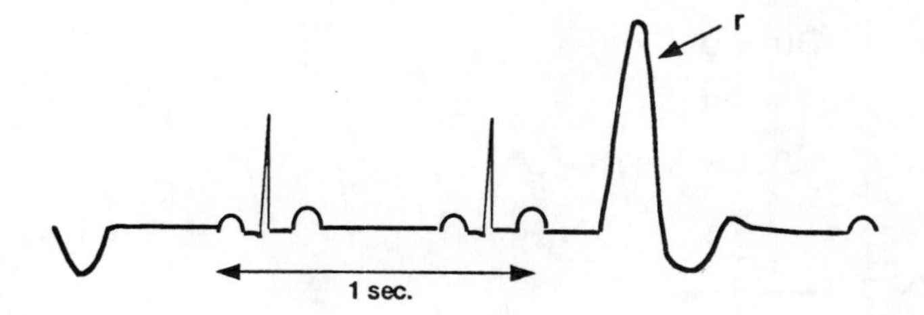

Figure 6–6. Electrocardiogram recorded from lead II in a patient.

141. Which of the following can be said regarding the ECG in Figure 6–6?

(A) it is an example of bigeminy
(B) the wave at point f is an R wave and will cause a stronger arterial pulse than the other R waves shown
(C) the deflection at point f will be associated with a premature ventricular systole
(D) A and B are correct
(E) A, B, and C are correct

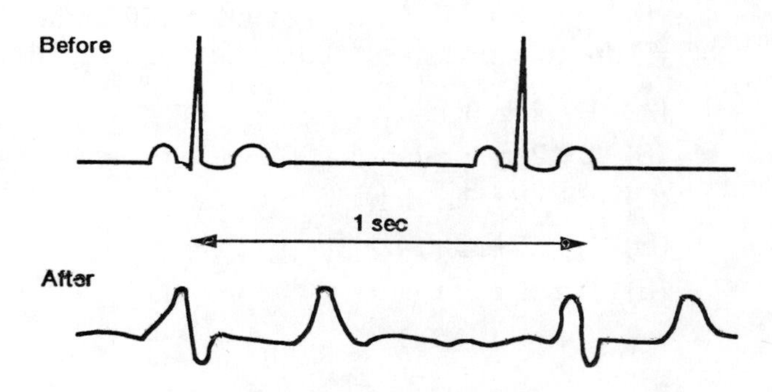

Figure 6–7. Electrocardiograms recorded from lead II in a patient after he developed anuria.

142. Which of the following statements best characterizes the patterns in Figure 6–7?

(A) ventricular fibrillation
(B) ventricular flutter
(C) ventricular tachycardia
(D) a slowing of cardiac conduction and an elevation of the T wave due to hyperkalemia
(E) a slowing of cardiac conduction and an elevation of the T wave due to hypokalemia

143. Which of the following conditions might produce hyperkalemia and be responsible for the abnormal ECG in Figure 6–7?

(A) hemolysis
(B) diabetes mellitus
(C) hyperaldosteronism
(D) A and B
(E) A and C

144. Which of the following statements is true regarding the records displayed in Figure 6–8?

(A) before therapy there was a sinus arrhythmia
(B) before therapy there was a periodic ventricular rhythm
(C) during therapy there was a conversion to ventricular tachycardia
(D) during therapy there was a conversion to a ventricular rhythm
(E) A and D

145. In the records seen in Figure 6–8, the drug produced

(A) a decreased pulse pressure
(B) an increased diastolic pressure
(C) a decreased heart rate, possibly due to a reflex stimulation of cardiac parasympathetic neurons
(D) a decreased heart rate, possibly due to a reflex inhibition of cardiac parasympathetic neurons
(E) B and C

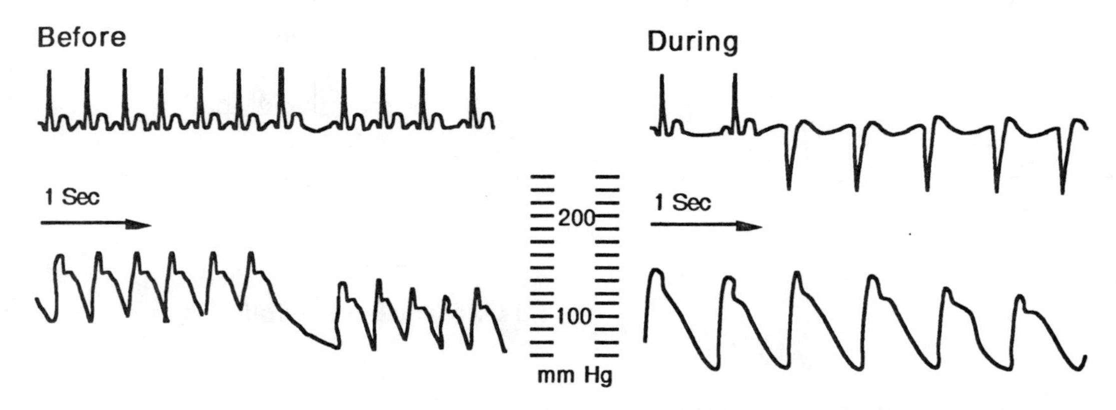

Figure 6–8. Electrocardiograms and blood pressure recording made before and during a patient's response to a drug.

146. In Figure 6–8, the mechanism responsible for a stimulation of cardiac parasympathetic neurons causing a ventricular rhythm is that parasympathetic neurons

- (A) act directly on the ventricles to decrease their rate of contraction
- (B) act directly on the ventricles to increase their irritability
- (C) act directly on the sinoatrial node to decrease its rate of firing
- (D) prevent atrial activation
- (E) block the atrioventricular node

147. Which of the following represents the major action of the drug given in the study in Question 144 (Fig. 6–8)?

- (A) vasoconstriction of the systemic arterioles
- (B) a decrease in total systemic peripheral resistance
- (C) vasoconstriction and a decrease in resistance
- (D) an increased sensitivity of the heart to acetylcholine
- (E) a decreased sensitivity of the heart to acetylcholine

148. Examining Figure 6–9, identify the following: (1) P wave of ECG, (2) second heart sound, (3) diastolic pressure, (4) dicrotic notch.

- (A) 1: j, 2: g, 3: l, 4: n
- (B) 1: j, 2: f, 3: l, 4: m
- (C) 1: k, 2: f, 3: l, 4: n
- (D) 1: g, 2: h, 3: f, 4: j
- (E) 1: g, 2: f, 3: i, 4: h

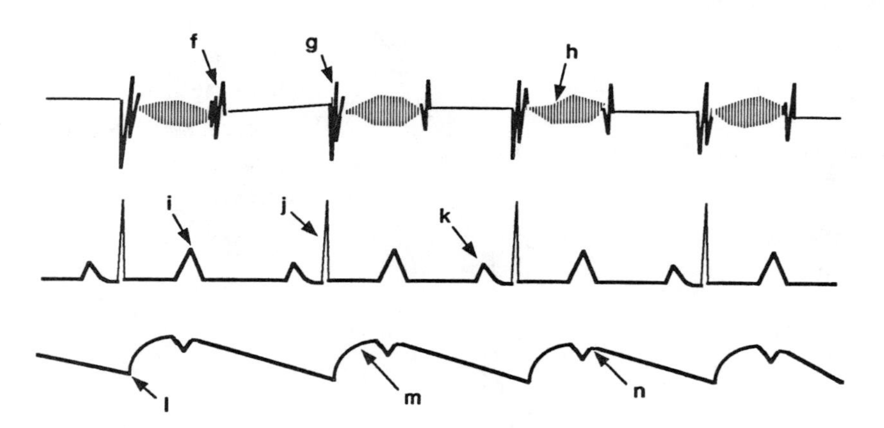

Figure 6–9. Phonocardiogram, electrocardiogram, and blood pressure recordings from a patient.

149. The transmembrane potential in cardiac tissue

(A) is independent of sodium ion concentration
(B) is primarily dependent on the tendency for potassium ions to leak out
(C) is primarily dependent on the tendency for sodium ions to leak in
(D) is normally positive inside with respect to the outside
(E) B and D

150. One can cause a cardiac Purkinje fiber to depolarize during phase 4 of the transmembrane potential by

(A) decreasing its Na^+ conductance
(B) decreasing its Ca^{++} conductance
(C) decreasing either its gNa^+ or its gCa^{++}
(D) decreasing its K^+ conductance
(E) decreasing its gNa^+, its gCa^{++}, or its gK^+

151. A pacemaker cell in the sinoatrial node differs from a follower cell in the ventricle in that the pacemaker cell has

(A) a smaller K^+ conductance during its diastolic potential
(B) a smaller diastolic potential
(C) A and B
(D) a phase 0 potential that is due primarily to Na^+ influx
(E) a steeper slope for its phase 0 potential

152. Which of the following statements is correct?

(A) K^+ gate opening initiates the fast current (in some cardiac muscle cells at a transmembrane [TM] potential of about –70 mv)
(B) Ca^{++} gate opening initiates the slow current (in some cardiac muscle cells at a TM potential of about –50 mv)
(C) K^+ gate closing is the most important cause of the repolarization (phase 3) of a cardiac muscle cell
(D) Na^+ gate opening is the action of acetylcholine on the cells of the SA node during the middle of ventricular diastole?
(E) B and D

Answers and Explanations

117. **(C)** Patterns g and h are both from follower cells. A pacemaker cell has an unstable resting transmembrane potential. The potential prior to the action potential in pattern h remains constant. Pacemaker cells slowly depolarize on their own in the absence of extrinsic stimuli. Follower cells, on the other hand, characteristically maintain a constant resting potential until an extrinsic stimulus arrives from a pacemaker cell or some other source.

118. **(A)** Once the threshold potential or firing level has been reached, there is a rapid depolarization of the cell. Point k represents the beginning of the cell's refractory period and the end of its irritability. The rheobase for a cell is the minimal stimulus strength in volts that will produce an action potential when that stimulus is applied for a prolonged period.

119. **(A)** Cardiac cells have a delay in repolarization that results in their having a prolonged period of inexcitability. The plateau type of potential is characteristic of the heart. The spike potential is characteristic of skeletal muscle fibers and axons.

120. **(A)** The heart's contraction is not initiated by a nerve impulse but by a cardiac cell that depolarizes in the absence of extrinsic stimuli. Characteristic of the heart is that it contains a single pacemaker area that dominates a number of follower cells and potential pacemaker cells. The impulse that the pacemaker originates is transmitted from one cell to the next.

121. **(C)** The SA node remains the pacemaker area even after its innervation is destroyed. The atria and ventricles contain numerous areas that will take over the pacing of the heart if the SA node is depressed, destroyed, or cut off from the rest of the heart. The cells of the sinoatrial node become the pacemaker for the heart because they stimulate other potential pacemaker cells before they are able to initiate their own action potentials.

122. **(E)** The atrium conducts at a rate of 0.9 m/sec. The ventricle conducts at a rate of from 0.3 to 1 m/sec. The Purkinje fibers conduct at a rate of 2 to 4 m/sec. The right bundle branch conducts at a rate of from 1 to 2 m/sec. The AV bundle, bundle branches, Purkinje fibers, and ventricular endocardium conduct action potentials rapidly and therefore cause an almost simultaneous activation of the fibers in the right and left ventricles. The AV junctional tissue at the AV node conducts at a rate of 0.05 m/sec. This very slow conduction helps delay the transmission of the impulse to the ventricles and therefore prevents the ventricles from contracting at the same time as the atria. All of these rates will vary with the degree of autonomic nervous system tone, the level of epinephrine in the blood, and other changes in the environment.

123. (B) Once the impulse has reached the apex, it travels up the endocardial surface of the right and left ventricles and moves from here to the epicardial surface. Because the left ventricular muscle mass is larger than the right, it takes longer to completely activate the left ventricle than the right. The last parts of the heart to develop action potentials are the base of the left ventricle and the superior interventricular septum. Apparently, the AV bundle and bundle branches in the basal portion of the interventricular septum are sufficiently isolated from their neighboring muscle fibers in the septum that the action potentials they contain pass by the rest of the septal muscle fibers without activating them, and it is not until the impulse comes around a second time that activation of the remaining septum occurs.

124. (A) In the absence of intercellular conduction, the heart would stop ejecting blood, the arterial pressure would decline, and there would be a reflex decrease in cardiac parasympathetic tone and an increase in sympathetic tone. When a potential pacemaker is isolated from its neighbors, it initiates its own action potentials. The development of multiple ectopic pacemakers leads to a chaotic type of contraction called fibrillation. While some fibers are contracting, others are relaxing. This loss of synchrony leads to such weak contractions that a pulse will not even be detected.

125. (C) The left leg lead goes to the positive terminal and the right arm lead to the negative terminal of the recorder. The positive terminal is the part of the recorder that causes a positive deflection when exposed to positivity and a negative deflection when exposed to negativity. Choice (A) is lead II. Choice (B) is lead I. Choice (D) is lead aVR. Choice (E) is lead aVL.

126. (D) Lead V2 is a precordial unipolar lead in which the exploratory electrode is at position 2 on the chest. Lead aVL is an augmented unipolar lead in which the equivalent to the exploratory electrode is on the left arm (a, augmented; V, vector; L, left arm). Standard limb lead I is a lead in which the potential between the left arm and right arm is measured.

127. (A) Lead aVR in the normal subject has inverted P, R, and T waves and an R wave of smaller amplitude than that found in lead II. Lead aVF normally has erect P, R, and T waves. Lead V4 normally has erect P, R, and T waves. Lead I in the normal subject characteristically has an R wave of lower amplitude than that found in lead II. In other words, this could be lead I if the subject had an electrical axis of 20°. An axis of 60 to 70° is much more likely.

128. (D) Plus 60° is approximately the electrical axis of the atria (ie, 60° is the direction of its frontal plane dipole). Marked deviations from this may be due to abnormal conduction patterns, the presence of an ectopic pacemaker, or an abnormal anatomical position for the heart.

129. (E) Lead I represents an electrical axis of 0°; lead II represents an electrical axis of 60°; lead III represents an electrical axis of 120°; and lead aVL represents an electrical axis of –30°.

130. (D) This is a period in the cardiac cycle when activation is spreading somewhat caudally from the endocardium to the epicardium of the right and left ventricles. If the major spread were from apex to base, one would expect an axis of approximately –60°. If the heart were vertical and the right and left ventricles were transmitting signals of equal strength with an axis of +120° (right ventricle) and +60° (left ventricle), respectively, one would expect an axis of +90°. The adult heart, however, is not vertical, and the left ventricle normally transmits a stronger signal.

131. (E) The dipole moment of the Q and S waves is considerably less than that of the P, R, and T waves. Their absence in certain leads is consistent with a healthy heart. It is particularly difficult to record an S wave in lead III, not only because of its low moment but also because of the relationship between the electrodes and dipole. The maximal deflection on the ECG is obtained when the dipole is moving toward one electrode and away from the other. When the electrodes are 90° out of phase with the dipole, they record no deflection. This indicates one advantage of recording more than one ECG lead.

132. **(C)** In a nodal rhythm, the ventricular conduction pattern is normal, but its relationship with the P wave is not. In a sinus rhythm, a P wave and an isoelectric line (ie, the PR segment) will precede each R wave. The R wave would be of longer duration and the T wave would usually be inverted. This is a heart rate of about 75/min, not between 250 and 350/min as in flutter. In fibrillation, the waves are of low amplitude (ie, below 0.2 mv).

133. **(B)** Parasympathetic neurons are capable of producing first-, second-, and third-degree block in response to increases in arterial pressure. Increased sympathetic tone to the AV node would decrease the PR segment. Increased sympathetic tone to the SA node speeds the heart under most conditions. Increased parasympathetic tone to the SA node slows the heart under most conditions. Decreased carotid sinus pressure would reflexly decrease parasympathetic tone and increase sympathetic tone.

134. **(D)** The QRS complex occurs at the beginning of the heart's refractory period. The QRS complex occurs during the depolarization of the ventricles. The PR interval occurs during atrial systole. The first heart sound begins during the QRS complex and ends prior to the beginning of the T wave.

135. **(B)** Parasympathetic tone to the SA node can be increased to such an extent that the SA node stops firing. When this happens, another pacemaker (ie, the AV node) takes over. In nodal rhythm there is a QRS complex of normal duration, and there may be retrograde atrial conduction or no apparent atrial conduction. In the case of the former, the P wave would be inverted and the PR segment shortened. In the case of the latter, there would be no PR segment.

136. **(A)** In most complete blocks the R-R′ interval exceeds 1.4 seconds, but one may obtain a pattern such as seen here when the block is associated with a marked cardiac sympathetic tone. In a second-degree heart block every second (2:1 block), or every third (3:1 block), etc, P wave is followed by a normal PR segment, QRS complex, and T wave. All other P waves are not followed by QRS and T waves. In sinus arrhythmia the P-P′ interval equals the R-R′ interval in each cardiac cycle, but they vary from one cycle to the next. In ventricular rhythm the duration of the QRS complex is more prolonged than above.

137. **(A)** One pacemaker is probably in or near the SA node and the second is in or near the AV node. The duration of the QRS complex is normal and, therefore, the velocity of conduction in the ventricles is also probably normal. The presence of well-defined P waves usually indicates the presence of atrial systole.

138. **(D)** The stimulation of cardiac sympathetic nerves causes a more forceful contraction of the ventricles. One sign of this is an increased rate of pressure development in the ventricles. Sympathetic neurons (1) act on the SA node to increase the heart rate (ie, decrease the R-R′ interval), (2 and 3) act on the atria and ventricles to increase their speed of conduction and decrease their refractory periods (decrease the duration of the ST segment and the PR segment), and (5) decrease the refractory periods of the AV node. By decreasing the refractory period of the AV node it is possible to prevent a second-degree heart block when the atrial rate is rapid.

139. **(B)** An early repolarization of the ventricular fibers would cause a decrease in the ST segment. Depression of the SA node would cause either an increase in the R-R′ interval or an ectopic rhythm. Slowing of conduction at the AV node would produce a prolongation of the PR segment. A conduction block in the left bundle branch would produce a prolongation of the QRS complex.

140. **(C)** (1) The isoelectric line is the pattern (g) that results from no potential between the two terminals of the recorder. When there is no injury potential, it lies between the P and QRS waves and between the QRS and T waves, and between the T and P waves. (2) The R wave (h) follows the pacemaker artifact (f). It is of abnormally long duration in the above pattern because the stimulating electrodes have been placed distal to the rapidly conducting fibers of the heart (ie, the AV bundle, its branches, the Purkinje fibers, and the endocardium). (3) The T wave (i) follows the R wave and, in this case, is partially incorporated into the R wave. In Figure 6.5 it is a negative deflection. This is characteristic of ventricular rhythm. The inverted T wave also occurs in cardiac ischemia. (4) The pacemaker artifact (f) is a short-duration, high-amplitude deflection. When the pacemaker electrodes are in the ventricle, it precedes the R wave. When they are in the atrium, it precedes the P wave.

141. **(E)** It is a trigeminy (one premature ventricular beat in every three cardiac cycles). The R wave at point f will cause a weaker than normal pulse. This is due in part to the reduced filling time for the heart and the less rapid activation for the ventricles. When one takes a pulse and finds every third wave weak, this pattern is also characterized as a trigeminy.

142. **(D)** Normally, the potassium concentration of the blood plasma is 5 mEq/L. An increase to 7 mEq/L will cause a prolonged P-R interval and QRS complex and an elevation of the T wave. The pattern in Figure 6.7 was produced by a concentration of potassium of 8 mEq/L of plasma. In ventricular fibrillation, the deflections of the ECG are more frequent and of lower amplitude than normal. In ventricular flutter and tachycardia, the cardiac cycle is of shorter duration.

143. **(D)** Aldosterone facilitates the loss of potassium from the body in the urine and sweat. It does not, either in the presence or absence of kidney malfunction, produce hyperkalemia. The K^+ is in higher concentration in the intracellular fluid than in the extracellular fluid. Degenerative conditions which result in the breakdown of the cell membrane or conditions which increase the permeability of the membrane can cause a marked and sometimes fatal movement of K^+ into the serum. This may result from hemolysis or the systemic acidosis produced in diabetes mellitus. It is most dangerous when the kidneys have either a lost or an impaired ability to excrete potassium.

144. **(E)** The presence of a P wave and a PR segment prior to each R wave indicates a sinus rhythm. The variability of the duration of the R-R′ interval is a form of arrhythmia, which is also sometimes called inspiratory tachycardia, since the R-R′ intervals shorten during each inspiration. Sinus arrhythmia (ie, inspiratory tachycardia) occurs in healthy individuals. Sinus arrhythmia is not due to a ventricular pacemaker. The therapy resulted in a heart rate of about 50/min. The long duration of the R wave is indicative of a ventricular pacemaker.

145. **(E)** The drug apparently caused an increase in diastolic, systolic, and pulsatile pressure in the systemic arteries. This, acting through the baroreceptors, caused a reflex stimulation of vagal parasympathetic neurons to the heart.

146. **(D)** The absence of P waves during the ventricular rhythm probably indicates that the parasympathetic neurons have prevented atrial activation, which releases a potential pacemaker in the ventricle from sinoatrial dominance. The parasympathetic neurons exert little direct action on the ventricles. Acetylcholine would act directly on the ventricles to decrease irritability. In heart block there might also be ventricular rhythm, but P waves would be seen if the block were restricted to the AV node.

147. **(A)** Arteriolar vasoconstriction, by reducing arterial drainage, produces a distention of the arteries that causes an increase in the arterial diastolic pressure. Vasoconstriction causes an increase in peripheral resistance. An increased sensitivity of the heart to acetylcholine would explain the cardiac slowing but not the increase in diastolic pressure. Cardiac slowing causes a decrease in diastolic pressure. What probably occurred during therapy is that a vasoconstrictor such as norepinephrine was injected and produced an increase in arterial pressure and a reflex slowing of the heart.

148. (C) 1. k. It normally precedes the R wave and is of shorter duration than the T wave.
2. f. It is heard after the closure of the semilunar valves, near the end of the T wave.
3. l. The pressure in the aorta that results from ventricular diastole (ie, the lowest pressure during any cardiac cycle) is called the diastolic pressure. It occurs during early ventricular ejection.
4. n. The dicrotic wave follows the dicrotic notch or incisura and occurs early in the catacrotic limb of the pressure curve. It should not be confused with a weak ventricular contraction.

149. (B) Transmembrane potential depends on the chemical and electrostatic forces across the cell membrane. This is due principally to the distribution of ions such as sodium, potassium, and calcium across the cell membrane. The difference in the permeability of the membrane to these ions results in a resting potential for that cell membrane.

150. (D) Depolarization is caused by an increase in Na^+ or Ca^{++} conductance or a decrease in K^+ conductance. (1) In skeletal muscle, acetylcholine causes depolarization by increasing gNa^+. (2) In the sinus node of the heart the pacemaker cell shows a phase 4 depolarization because of its low gK^+. (3) Norepinephrine increases the heart rate because it increases gCa^{++} in the pacemaker cell during phase 4.

151. (C) The smaller K^+ conductance, by moving the SA node cell's transmembrane potential away from the equilibrium potential for K^+, decreases the diastolic potential, eliminates the fast Na^+ current during phase 0, and produces a phase 0 that is caused by a slow Ca^{++} current. This slow current causes the lower slope during phase 0 (ie, a lower dV/dt: rate of change of voltage).

152. (B) Na^+ gate opening initiates the fast current, while Ca^{++} gate opening initiates the slow current. K^+ gate opening is responsible for repolarization or hyperpolarization of the membrane. Closure or partial closure of the K^+ gate moves the TM potential away from the equilibrium potential for K^+.

CHAPTER 7

The Heart–Mechanical Properties

Questions

DIRECTIONS (Questions 153 through 194): Each of the numbered items or incomplete statements in this section is followed by answers or by completions of the statement. Select the ONE lettered answer or completion that is BEST in each case.

153. Which of the following statements regarding the pressure pattern seen in Figure 7–1 is correct?

(A) the catheter was in the right ventricle
(B) the P wave of the electrocardiogram begins immediately prior to j
(C) the period of most rapid ventricular filling begins at h
(D) B and C are correct
(E) all are correct

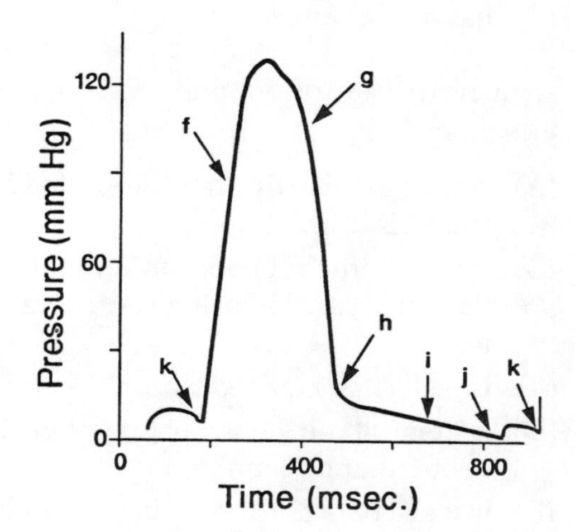

Figure 7–1. Record through a catheter from a patient with normal cardiovascular function.

154. In a resting, healthy individual, the pulmonary valve opens when the pressure in the right ventricle is approximately

(A) 10 mm Hg
(B) 30 mm Hg
(C) 50 mm Hg
(D) 80 mm Hg
(E) 120 mm Hg

155. The second component of the first heart sound is most closely associated with what event during the cardiac cycle?

(A) opening of the mitral valve
(B) closure of the mitral valve
(C) closure of the aortic valve
(D) closure of the pulmonary valve
(E) closure of the tricuspid valve

156. The second component of the second heart sound is usually most clearly heard on the ventral surface of the chest at the

(A) second intercostal space to the right (patient's right) of the sternum
(B) second intercostal space to the left of the sternum
(C) fifth intercostal space to the left of the sternum
(D) fifth intercostal space at the sternum
(E) ventral tip of the xiphoid process

157. The first component of the first heart sound is usually most clearly heard on the ventral surface of the chest at the

(A) ventral tip of the xiphoid process
(B) fifth intercostal space over the sternum
(C) fifth intercostal space to the left (patient's left) of the sternum
(D) second intercostal space to the left of the sternum
(E) second intercostal space to the right of the sternum

158. A systolic murmur was heard over the manubrium of the sternum. The most likely diagnosis is

(A) an increased hematocrit
(B) aortic stenosis
(C) aortic insufficiency
(D) mitral stenosis
(E) patent ductus arteriosus

159. Systolic and diastolic murmurs are heard over the body of the sternum. The most likely diagnosis is

(A) aortic and pulmonary stenosis
(B) aortic and pulmonary insufficiency
(C) aortic insufficiency and stenosis
(D) an increased hematocrit
(E) a decreased hematocrit

160. A continuous murmur is heard over the manubrium of the sternum. The most likely diagnosis is

(A) an increased hematocrit
(B) aortic stenosis
(C) aortic insufficiency
(D) mitral stenosis
(E) patent ductus arteriosus

161. Which of the following statements is true regarding turbulence?

(A) it occurs in the normal cardiovascular system during ventricular ejection, particularly at elevated cardiac output
(B) it always occurs when the Reynolds number exceeds 1000 (equation using r)
(C) its development is influenced only by the variables embodied in the Reynolds equation
(D) A and C
(E) all are correct

162. Turbulent blood flow in human subjects results from a decreased

(A) hematocrit
(B) critical Reynolds number (this occurs in a saccular aneurysm)
(C) hematocrit or critical Reynolds number
(D) cardiac output
(E) cardiac output or hematocrit

163. A physician notes that when taking a patient's arterial blood pressure using a sphygmomanometer cuff, although the sound over the artery becomes muffled at a cuff pressure of 70 mm Hg, it does not disappear as the cuff pressure returns to 0 mm Hg. What conclusion is most likely drawn from these data? The patient

(A) has a diastolic pressure at or near 0 mm Hg (ie, may have aortic insufficiency)
(B) probably has a weakened heart (ie, left ventricular congestive heart failure)
(C) has a patent ductus arteriosus
(D) has an aortic stenosis
(E) has a low hematocrit

164. At a heart rate of 80/min, coronary artery blood flow

(A) is greatest shortly after the second heart sound is heard
(B) is zero in the subendocardial portion of the left ventricle during ventricular systole
(C) is well characterized by both A and B
(D) is greatest during the period of peak left ventricular ejection
(E) is well characterized by both A and D

165. Which of the following findings is indicative of coronary ischemia?

(A) a higher lactate concentration in the coronary sinus than in the aorta
(B) an increased coronary AVO_2 concentration difference
(C) both of the above
(D) an increased coronary blood flow
(E) a decreased coronary blood flow

166. Which of the following is the most common cause of an increased coronary blood flow?

(A) a decreased coronary perfusion pressure
(B) a decreased stimulation of β_2 adrenergic receptors in the heart
(C) an increased stimulation of α adrenergic receptors in the heart
(D) an increased stimulation of β_1 adrenergic receptors in the heart
(E) an increased O_2 concentration in the coronary arteries

167. The principal energy source for the heart of a healthy fasting subject is

(A) free fatty acids
(B) glucose
(C) lactate and pyruvate
(D) amino acids
(E) polypeptides

168. Which of the following statements best characterizes the ductus venosus of the fetus?

(A) it receives most of its blood from the umbilical vein
(B) it sends its blood directly into the inferior vena cava
(C) A and B
(D) it contains blood with a higher O_2 content than the blood in the abdominal aorta of the fetus
(E) all of the above statements are correct

169. Which of the following is true within a few minutes after the birth of a healthy child?

(A) permanent fusion of the septum primum and secundum is complete
(B) the foramen ovale is anatomically closed
(C) A and B
(D) the umbilical arteries are functionally closed by contraction of the smooth muscles in their walls
(E) B and D

170. Which of the following cardiovascular changes is characteristic of the transition from the fetal to the newborn state?

(A) left and right ventricles change from a series to a parallel circulation
(B) the pressure in the aorta rises until it exceeds the pressure in the pulmonary artery
(C) right atrial pressure becomes greater than left atrial pressure
(D) circulatory resistance through the lungs increases
(E) the left heart begins to pump more blood than the right heart

171. A patient is diagnosed as having an opening in the interatrial septum uncomplicated by other cardiac abnormalities. Which of the following would best confirm this diagnosis?

(A) a systolic murmur
(B) an elevated PO_2 in the pulmonary artery
(C) a decreased pressure in the right atrium
(D) an elevated pressure in the left atrium
(E) cyanosis

172. During a cardiac catheterization, the following data are collected from two 50-year-old men:

	Patient 1	Patient 2
Aorta:		
Pressure (mm Hg)	140/50	95/65
O_2 saturation (%)	95	96
Left ventricle:		
Pressure (mm Hg)	140/15	95/9
O_2 saturation (%)	95	96
Left atrium:		
Pressure (mm Hg)	12	31
O_2 saturation (%)	95	96
Pulmonary artery:		
Pressure (mm Hg)	26/11	68/21
O_2 saturation (%)	75	74
Right ventricle:		
Pressure (mm Hg)	26/5	68/7
O_2 saturation (%)	75	74
Right atrium:		
Pressure (mm Hg)	5	6
O_2 saturation (%)	75	74

Select the one best diagnosis below for patient 1.

(A) aortic insufficiency associated with left ventricular hypertrophy
(B) aortic stenosis associated with left ventricular hypertrophy
(C) tricuspid insufficiency
(D) patent ductus arteriosus
(E) systemic hypertension due to arteriolar constriction

173. Select the one best diagnosis below for patient 2 in question 172.

(A) aortic insufficiency
(B) aortic stenosis
(C) tricuspid insufficiency
(D) tricuspid stenosis
(E) mitral stenosis

174. The following records (Fig. 7–2, A through E) were obtained from five patients. Each record contains a left atrial, left ventricular, and aortic pressure tracing. The paper speed, 0 baseline, and amplification for all parameters in each study are the same. Which tracings refer to the following conditions?

(A) A = aortic stenosis, E = mitral stenosis
(B) B = normal, D = mitral insufficiency
(C) A = aortic insufficiency, C = aortic stenosis
(D) D = mitral insufficiency, E = aortic stenosis
(E) B = mitral stenosis, C = aortic insufficiency

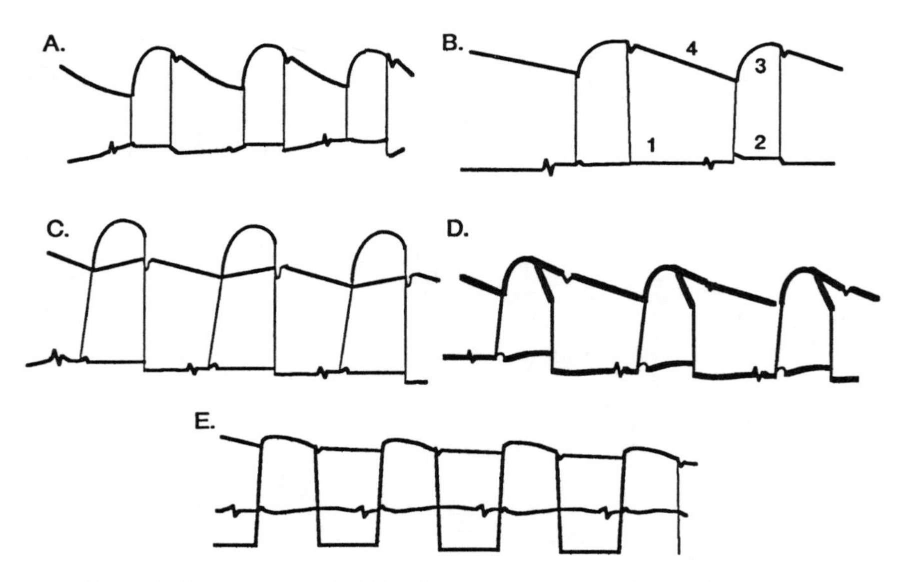

Figure 7–2. Record of left atrial, left ventricular, and arterial blood pressures.

175. The following data are collected from a patient:

Respiratory tidal volume	230 mL
O_2 consumption	110 mL/min
Femoral artery O_2	20 mL/dL ($PO_2 = 100$ mm Hg)
Femoral vein O_2	13 mL/dL ($PO_2 = 35$ mm Hg)
Pulmonary artery O_2	14 mL/dL ($PO_2 = 40$ mm Hg)

What would the cardiac output of this patient be?

(A) less than 1600 mL/min
(B) between 1600 and 1900 mL/min
(C) between 1900 and 2200 mL/min
(D) between 2200 and 2500 mL/min
(E) more than 2500 mL/min

176. The following data were collected in the right heart (ventricle and pulmonary artery) and the left heart (ventricle and aorta):

	Right	**Left**
Ventricular pressure (mm Hg)	70/8	100/8
Arterial pressure (mm Hg)	70/20	100/70

On the basis of these data, what conclusions can you draw?

(A) right ventricular systolic pressure is higher than normal
(B) pulmonary arterial pressure is lower than normal
(C) A and B are correct
(D) the abnormal data may be due to tricuspid stenosis
(E) A and D are correct

177. Within limits, an increase in the end-diastolic volume of the healthy right ventricle will usually

(A) increase the mean ejection pressure of the ventricle
(B) increase the stroke work of the ventricle
(C) A and B are correct
(D) decrease the stroke volume of the ventricle
(E) B and D are correct

178. When an average, healthy, 20-year-old person changes from a relaxed standing position to running with a maximal effort, he/she is capable of

(A) increasing stroke volume up to twofold
(B) increasing stroke volume up to fourfold
(C) increasing stroke volume up to sixfold
(D) increasing stroke volume up to ninefold
(E) increasing stroke volume up to twelvefold

179. The following data are collected from a resting patient before and after surgery:

	Before	**After**
Diameter of the lumen of the left ventricle (cm)	8	10
Heart rate (cycles/min)	70	90
Peak pressure in left ventricle (mm Hg)	110	100

Assuming that the ventricle is a sphere, calculate the tension the left ventricle has to develop before and after surgery in order to produce the pressures indicated. Which of the following best summarizes your conclusions? The surgery causes

(A) a 10 to 20% reduction in the tension exerted by the left ventricle
(B) a 10 to 20% increase in the tension exerted by the left ventricle
(C) little or no change in the tension exerted by the left ventricle
(D) more than a 20% reduction in the tension exerted by the left ventricle
(E) more than a 20% increase in the tension exerted by the left ventricle

180. Which of the following will cause an increase in left ventricular preload?

(A) an increased end-diastolic volume
(B) an increased end-diastolic pressure
(C) both of the above
(D) an increased end-systolic radius
(E) an increased stroke volume

181. Which of the following is a response of the heart to the stimulation of adrenergic sympathetic neurons?

(A) it shifts the Starling curve of the heart to the left
(B) it increases conduction time
(C) A and B are correct
(D) it increases the refractory period
(E) A and D are correct

182. Which of the following factors is a response of the healthy heart to the stimulation of cardiac parasympathetic neurons?

(A) a shift of the Starling curve for the atria to the right
(B) a shift of the Starling curve for the ventricles to the left
(C) A and B are correct
(D) a speeding of conduction at the AV node
(E) all are correct

183. Which of the following statements is most accurate?

(A) Ca^{++} and K^{+} have synergistic actions on the heart
(B) Ca^{++} has a positive inotropic action on the heart
(C) K^{+} has a positive inotropic action on the heart
(D) Ca^{++} and K^{+} have a positive inotropic action on the heart and act synergistically
(E) the intravenous injection of KCl causes hyperpolarization of cardiac cells

184. Which of the statements concerning the left ventricle of a resting adult is true?

(A) it has a stroke volume equal to that of the right ventricle
(B) its stroke work is four to five times greater than that of the right ventricle
(C) A and B are correct
(D) its stroke work index is equal to that of the right ventricle
(E) all are correct

185. Which of the following results from an increase in the inotropicity of the left ventricle?

(A) an increase in its ejection fraction
(B) an increase in its pre-ejection period/ejection time ratio
(C) A and B are correct
(D) an increase in the end-systolic volume of the ventricle
(E) a decrease in the ventricle's V_{max}

186. A patient has the following changes in response to therapy:

	Before	**After**
Mean arterial pressure (mm Hg)	100	70
Cardiac output (L/min)	5	4
Heart rate (beats/min)	70	90
End-diastolic volume of left ventricle (mL)	140	240

Which of the following statements about these data is true? The therapy produced

(A) an increased left ventricular stroke work
(B) a negative inotropic response
(C) A and B
(D) a decreased total systemic peripheral resistance
(E) B and D

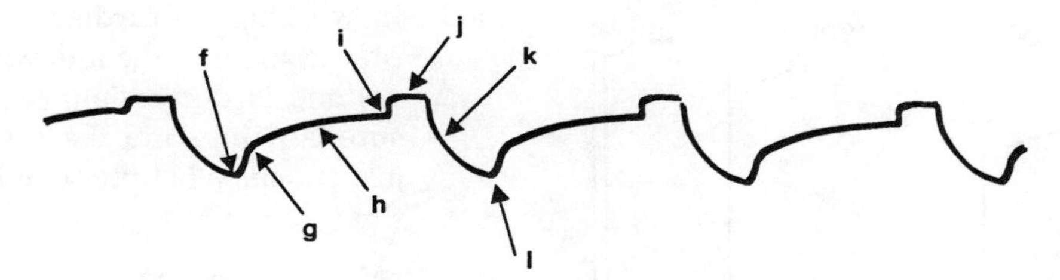

Figure 7–3. Tracing of a patient's left ventricular volume during several cardiac cycles.

187. With reference to Figure 7–3, which of the following is correct?

(A) isovolumic relaxation occurs at g
(B) atrial systole occurs at i
(C) mitral valve opens at h
(D) isovolumic systole occurs at k
(E) diastasis occurs at j

188. With reference to Figure 7–4, which of the following is correct?

(A) M represents closure of the mitral valve
(B) N represents opening of the mitral valve
(C) L represents closure of the aortic valve
(D) O–L represents ventricular filling
(E) L–M represents isovolumic ventricular contraction

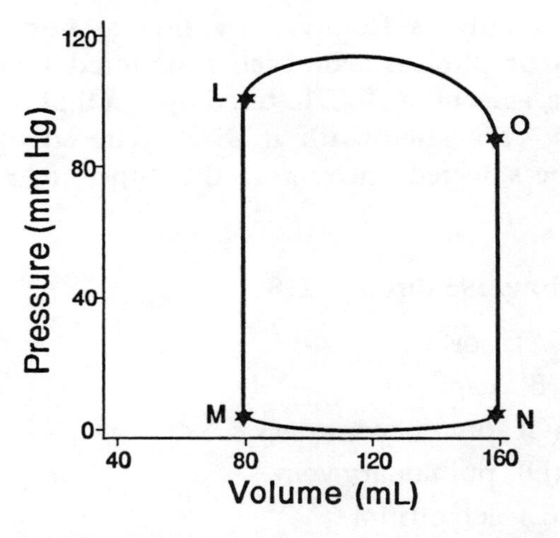

Figure 7–4. Pressure–volume loop recorded from the left ventricle.

189. Which of the following is true regarding mitral valve motion?

(A) the posterior leaflet is the most mobile
(B) the valve annulus area decreases during valve closure
(C) papillary muscle contraction plays no role in valve closure
(D) closure is entirely due to the increase in ventricular pressure
(E) the valve leaflets remain far apart throughout ventricular diastole

190. The pressure–volume loops in Figure 7–5 show which of the following changes from BEFORE to DURING medication?

(A) a decrease in the total systemic resistance
(B) an increased stroke volume caused by a decreased ventricular afterload
(C) A and B are correct
(D) an increased preload
(E) B and D are correct

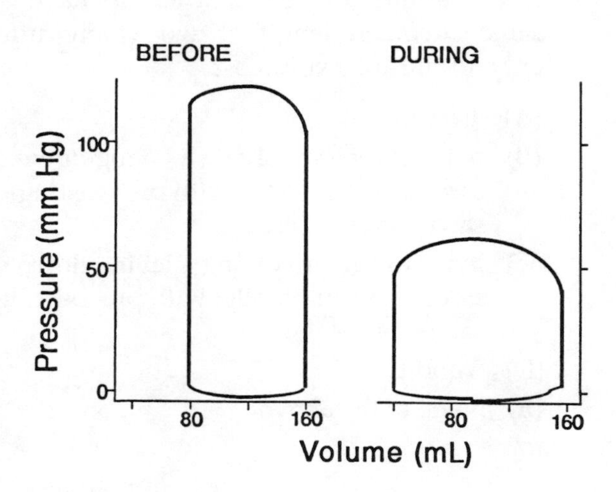

Figure 7–5. Pressure–volume loops recorded from a patient's left ventricle before and during medication.

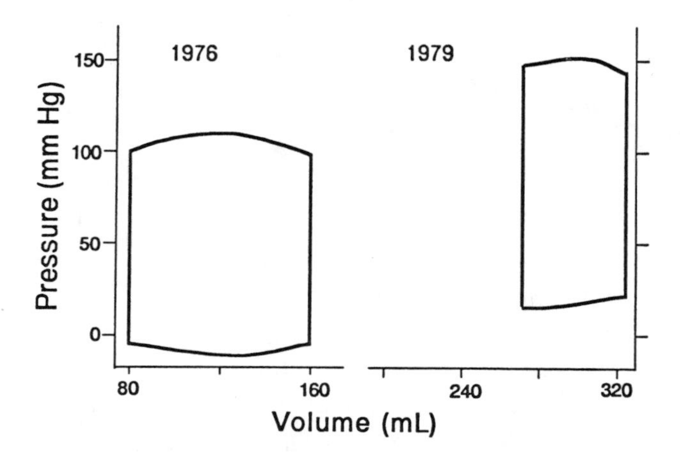

Figure 7–6. Pressure–volume loops obtained from the left ventricle of a subject in 1976 and 1979.

191. Regarding Figure 7–6, which of the following are consistent with the diagnosis below?

(A) improved the function of his heart through a demanding exercise regime
(B) developed a Frank–Starling curve that has shifted to the left
(C) increased his stroke work by a Frank–Starling mechanism
(D) B and C are correct
(E) developed a left ventricular failure in response to an aortic stenosis

192. In the study described in Figure 7–6, the end-diastolic left ventricular sarcomere length was 2.1 µm in 1976 and 1979. What mechanism permits the left ventricle to have the same sarcomere length at two widely different end-diastolic volumes?

(A) atrophy
(B) hypertrophy involving a laying down of sarcomeres in series with pre-existing sarcomeres
(C) hypertrophy involving a laying down of sarcomeres in parallel with pre-existing sarcomeres
(D) B and C
(E) None of the above

193. In working up cardiac data on a patient, a physician notes the following: left ventricular volume is decreasing, left ventricular pressure is rising, and the first heart sound has just ended. What phase of the cardiac cycle is this?

(A) early ejection
(B) mid-ejection
(C) late ejection
(D) rapid filling
(E) isovolumic ventricular diastole

194. Select the one best statement:

(A) an acute increase in the ventricular end-diastolic volume causes an increased ventricular efficiency
(B) the three major determinants of myocardial oxygen consumption are (1) cardiac afterload, (2) inotropic state of the heart, and (3) heart rate
(C) both of the above are true
(D) an aortic stenosis associated with a normal mean arterial pressure and cardiac output can markedly increase myocardial O_2 consumption
(E) all of the above

DIRECTIONS (Questions 195 through 201): Each set of questions in this section consists of a list of lettered options followed by several numbered words or phrases. For each numbered word or phrase, select the ONE lettered option that is most closely associated with it. Each lettered option may be selected once, more than once, or not at all.

Questions 195 through 198

(A) aorta
(B) capillary
(C) pulmonary artery
(D) pulmonary vein
(E) left atrium
(F) left ventricle
(G) precapillary sphincter

(H) right atrium
(I) right coronary artery
(J) right ventricle
(K) systemic arteriole
(L) systemic artery
(M) veins
(N) venule

195. Using an indwelling catheter, a blood sample and blood pressure measurements are made on a normal patient, giving values of 73% saturation and 28/15 mm Hg, respectively. At what point in the circulation is the catheter tip drawing the samples/making the measurements?

196. Using an indwelling catheter, a blood sample and blood pressure measurements are made on a normal patient, giving values of 97% saturation and 124/8 mm Hg, respectively. At what point in the circulation is the catheter tip drawing the samples/making the measurements?

197. What in the circulation holds most of the body's blood volume?

198. What site in the circulation shows the greatest Windkessel effect, ie, hydraulic filtering?

Questions 199 through 201

(A) rapid filling
(B) diastasis
(C) isovolumic ventricular contraction
(D) isovolumic ventricular relaxation
(E) early ejection
(F) mid-ejection
(G) late ejection
(H) atrial contraction
(I) dicrotic notch
(J) aortic valve opening
(K) mitral valve opening
(L) pulmonic valve closure
(M) mitral valve closure

199. Event that signals the time at which aortic diastolic pressure has been reached

200. Left ventricular pressure, while low, is rising slowly, and the associated ECG interval is isoelectric

201. An event resulting from the backward movement of a pressure wave through the aorta

Answers and Explanations

153. (D) The peak pressure in the right ventricle of the resting, healthy subject averages about 27 mm Hg, not 120 mm Hg as seen in Figure 7.1. The interval j–k represents atrial systole and is preceded by depolarization of the atrium. Point g represents the closing of the aortic valve and the end of ventricular systole. Point h represents the opening of the mitral valve and the beginning of rapid filling.

154. (A) 10 mm Hg is the diastolic pressure of the pulmonary artery. In other words, this is the minimum pressure in the pulmonary artery during a single cardiac cycle. 30 mm Hg is the systolic pressure of the pulmonary artery. In other words, it is the maximum pressure during a single cycle. 80 mm Hg is the pressure at which the aortic valve opens. 120 mm Hg is the systolic pressure in the aorta.

155. (E) Because right ventricular activation lags behind left ventricular activation by a few milliseconds, the increase in pressure that helps close the tricuspid valve and sets up the vibrations that produce the tricuspid component of the first sound lags behind the mitral component of the first sound. The opening of the mitral valve is followed by the third heart sound. The closure of the mitral valve is associated with the first component of the first heart sound. The closure of the aortic valve is associated with the first component of the second heart sound. The closure of the pulmonary valve is associated with the second component of the second heart sound.

156. (B) The second intercostal space to the right of the sternum is where the first component of the second sound is best heard. The second heart sound, for the most part, is due to vibrations of the walls of the ascending aorta and pulmonary artery. The pulmonary artery, as it leaves the base of the right ventricle, crosses to the left of the ascending aorta. The second heart sound, which is due to vibrations of the pulmonary artery (second component), is best heard to the left of the point where the first component of the second heart sound is best heard.

157. (C) The first heart sound is due to vibrations from the ventricles. The two ventricles lie cephalad to the tip of the xiphoid process. The right ventricle lies beneath the sternum at the fifth intercostal space. It is here that the second component of the first heart sound is best heard. It is at the apex of the heart where the first component of the first heart sound is usually most distinctly heard. It is the vibrations from the left ventricle that are responsible for this sound.

158. (B) If, during ventricular ejection, the blood from the left ventricle must be pumped through a narrow lumen, a high-velocity jet of blood will be pushed into the aorta, and turbulence will result. When the aortic valve closes and ejection ceases, the turbulence will also cease. The turbulence may be either pansystolic or midsystolic but will not occur during diastole unless there is some other disturbance in cardiac function.

159. (C) Aortic and pulmonary stenosis causes only a systolic murmur. Aortic and pulmonary insufficiency causes only a diastolic murmur. The aortic insufficiency causes a murmur as the blood rushes into the relaxing left ventricle from the aorta. The stenosis causes a murmur when the blood is being ejected into the aorta from the left ventricle. A decreased hematocrit is more likely to produce a continuous murmur. It never causes systolic and diastolic murmurs.

160. (E) A patent ductus arteriosus in the adult is a vessel through which blood usually moves from the aorta to the pulmonary artery (ie, a left-to-right shunt). Because the pressure in the aorta is considerably higher than in the pulmonary artery, there is a high-velocity flow. A sign of a high-velocity flow is turbulence (ie, a murmur). A murmur (ie, turbulent blood flow) is caused by a decreased hematocrit. This causes a systolic murmur. Aortic insufficiency and mitral stenosis cause a diastolic murmur.

161. (A) The Reynolds equation does not contain all of the variables that influence the development of turbulence; thus, under certain circumstances, the Reynolds number may well exceed 1000 before turbulence begins.

162. (C) A decrease in the hematocrit causes a decreased blood viscosity. Turbulence is frequently heard in the area of a saccular aneurysm. Increases in volume flow cause turbulence if the increases are sufficiently great.

163. (E) When taking an arterial pressure by this method, you have two indices of true diastolic pressure: (1) the pressure in the cuff at which a muffling of the sound is heard through the stethoscope, and (2) the pressure at which the disappearance of the sound is noted. Whenever this method yields widely divergent values for diastolic pressure, the muffling is the best index of arterial diastolic pressure. In other words, this patient does not have a diastolic pressure that would indicate either aortic insufficiency or left ventricular failure. One would need more evidence to come to either of these conclusions. A patent ductus arteriosus and an aortic stenosis produce a murmur that can be heard on or near the surface of the sternum but will not be heard in the forearm, where pressure is usually obtained. As the hematocrit decreases, the viscosity of the blood decreases and murmurs are more likely. The failure of the sound to disappear when one takes a pressure reading by the cuff method may be due to this continuous background murmur.

164. (C) Prior to the second heart sound, there is ventricular systole. During this period, ventricular contraction is occurring and causing the occlusion of the subendocardial coronary vessels. During ventricular diastole, the aortic pressure is decreasing and the coronary vessels are becoming more patent. The net effect is an increase in coronary blood flow during ventricular relaxation. The events noted in choices (B) through (E) all occur during ventricular systole, when many of the coronary vessels are occluded by the contracting cardiac muscle fibers.

165. (C) Normally, the heart catabolizes more lactic acid than it produces. During hypoxia the heart becomes more dependent on the anaerobic production of pyruvate and lactate from glucose as a source of energy, and the coronary sinus lactate concentration goes up, as does the $A\dot{V}O_2$ difference. Changes in coronary blood flow are usually a sign of a dynamic system of vessels capable of changing its resistance in response to the changing demands of the organ being served. Coronary ischemia occurs when the vessels of the heart lose their ability to maintain a level of O_2 delivery that permits the heart to catabolize more lactate than it produces.

166. (D) An increased perfusion pressure will cause an increased flow if there is not an associated increase in resistance. In the heart, when the perfusion pressure is maximum, many of the coronary vessels are being constricted by the contracting ventricular fibers. The stimulation of β_1 cardiac receptors causes (1) a positive chronotropic and inotropic cardiac response, (2) an increased cardiac metabolism, (3) a decreased PO_2 and the accumulation of metabolites, (4) coronary arteriolar dilation in response to this change in environment, and (5) an increased coronary flow. Item (3) is the usual cause of a coronary vasodilation. Other apparently less important causes are stimulation of β_2 receptors in the coronary arterioles and a decreased stimulation of α receptors in the coronary arterioles.

167. **(A)** The catabolism of free fatty acids in the fasting subject is responsible for about 65% of the heart's O_2 consumption; the catabolism of glucose, 18%; the catabolism of lactate, 16%. These figures will change when the concentration of nutrients in the blood changes. The catabolism of lactate, for example, will increase in strenuous exercise. The catabolism of glucose will decrease in diabetes mellitus and increase in response to insulin. Skeletal muscle and the brain, on the other hand, have a much greater preference for glucose catabolism as a source of energy than does the heart.

168. **(E)** The ductus venosus carries O_2 and other nutrients from the umbilical vein to the inferior vena cava, where it is mixed with less well-oxygenated blood. The blood in the ductus venosus is about 80% saturated with O_2 and that in the fetal aorta is about 60% saturated.

169. **(D)** The septum primum and secundum separate the right and left atria and form the foramen ovale. Their permanent fusion will not be complete until 2 to 8 weeks after birth. The foramen ovale lies in the interatrial septum, and the ductus arteriosus connects the pulmonary artery and aorta. Prior to birth, they provide a right-to-left shunting of the blood past the lungs. Immediately after birth, there is a fivefold increase in lung volume after the first breath is taken and a five- to tenfold increase in pulmonary blood flow due in part to the closure of the foramen ovale and ductus arteriosus. The umbilical arteries carry blood from the common iliac artery of the fetus toward the intervillous space of the placenta.

170. **(B)** When the placental circulation is cut off, the systemic vascular resistance rises. The closure of the ductus arteriosus and the foramen ovale converts a parallel circulation to a series circulation. It is the reversal of the pressure head in the atria (increase in left atrial pressure and decrease in right atrial pressure) during birth that is responsible for the closure of the foramen ovale. The asphyxia associated with birth facilitates gasping movements in the fetus, which will usually produce an intrapleural pressure between –30 and –50 mm Hg. At the same time, the pulmonary vascular resistance through the lungs becomes less than 20% of what it was prior to labor. This is, in part, responsible for a marked increase in blood flow through the pulmonary arteries, capillaries, and veins. The two ventricles eventually come to have the same output.

171. **(B)** The right atrial PO_2 is normally markedly lower than the left atrial PO_2, and therefore the left-to-right shunt of blood through the interatrial septum would markedly increase the right atrial PO_2. The pressure differences between the right and left atria are small and are therefore unlikely to cause a sufficiently high-velocity flow to produce a murmur. Cyanosis is neither a specific indication nor likely in this condition.

172. **(A)** The aortic diastolic pressure is about 30 mm Hg lower than normal, and the left ventricular end-diastolic pressure is 5 to 10 mm Hg higher than normal. This is characteristic of aortic insufficiency, a condition in which there is a leakage of blood from the aorta back into the ventricle during ventricular diastole. This results in the decrease in aortic diastolic pressure (ie, increased diastolic runoff of pressure and blood) and the increase in the ventricular end-diastolic pressure (ie, ventricular distention). The aortic pulse pressure in this patient ($140 - 50 = 90$ mm Hg) is also quite high. This is probably in part due to the increased end-diastolic volume of the ventricle, which causes an increased stroke volume through a Starling mechanism and through ventricular hypertrophy.

Aortic stenosis is a condition in which there is an increased resistance to left ventricular ejection. This results in a markedly lower aortic systolic pressure than is found in the left ventricle. Since the systolic pressures in the left ventricle and the aorta are the same, it is unlikely that the patient has aortic stenosis. In tricuspid insufficiency, an elevated right atrial pressure would be expected. In a patent ductus arteriosus, an elevated pressure in the pulmonary artery would be expected. Arteriolar constriction produces an increase in the arterial diastolic pressure as well as in the mean arterial pressure.

173. **(E)** The pressures in the chambers carrying blood to the left ventricle (left atrium, pulmonary artery, and right ventricle) are all elevated. The pressures in the left ventricle and aorta, on the other hand, are reduced. Normally, during ventricular diastole, left atrial and ventricular pressures are similar (7 vs. 7 mm Hg, for example). In this case, they are not (31 vs. 9 mm Hg). Therefore, there is an area of high resistance during ventricular diastole between the left atrium and left ventricle (ie, mitral stenosis). In other words, an area of high resistance is an area where there is a large pressure drop:

[resistance = (pressure #1 – pressure #2)/flow].

174. **(C)** The lower tracing on each record is from the left atrium. The upper tracing is from the aorta. The connecting tracing is from the left ventricle. A stenosis is an area of abnormally high resistance. Because resistance is directly related to perfusion pressure (P1 – P2), an abnormally high pressure difference across the stenosis results. In aortic stenosis, the abnormally high perfusion pressure exists between the left ventricle and aorta during ventricular systole.

An insufficiency is an area of abnormally low resistance and therefore a reduced P1 – P2. In aortic insufficiency, the aortic valve does not close completely and therefore there is a regurgitation of blood back into the ventricle and a reduced P1 – P2 between the ventricle and aorta during diastole. Note that the reduced pressure head is most marked at the end of diastole.

In the normal heart there is (1) practically no resistance between the atrium and ventricle during most of ventricular diastole (period of rapid filling through period of atrial systole), (2) an infinite resistance between the atrium and ventricle during ventricular systole, (3) practically no resistance between the ventricle and aorta during ventricular ejection, and (4) an infinite resistance between the ventricle and aorta during all of ventricular diastole. In the resting subject, the aortic pulse pressure is usually about one third the systolic pressure, not over one half as seen in record A (aortic insufficiency).

In mitral insufficiency the reduced pressure head is between the atrium and ventricle during ventricular systole. Note that this reduction is most marked at the end of ventricular systole. The increased pressure head is between the atrium and ventricle during ventricular diastole. The low aortic pulse pressure is also seen in aortic stenosis and when ventricular contractions are abnormally weak. If aortic stenosis is associated with arteriosclerosis, the aortic pulse pressure could be normal.

175. **(B)** Cardiac output = O_2 consumption/arteriovenous O_2 difference = 110 mL O_2/min/(20 – 14) (mL O2/dL blood) = 1833 mL/min

In this calculation, one must use the O_2 concentration in the pulmonary artery, because this value is more representative of the average for all the systemic veins than is the femoral vein value. Because O_2 is not removed from the blood between the left heart and the systemic capillaries, the O_2 concentration in all systemic arteries should be the same. The partial pressures of O_2 in mm Hg cannot be used in this calculation. They are representative of the plasma concentration of O_2, not the total blood content.

176. **(A)** Right ventricular and pulmonary artery systolic pressures are higher than normal. Tricuspid stenosis would not cause the conditions observed.

177. **(C)** Starling stated: "The energy of contraction is a function of the length of the muscle fiber." We can demonstrate in a heart–lung preparation that, within limits, an increase in the end-diastolic volume of a chamber of the heart causes an increase in the product of stroke volume and mean ejection pressure for that chamber. This product is called the stroke work. The stroke work index is the stroke work divided by the surface area of the subject. The surface area of the average adult man is approximately 1.7 m^2. The stroke work performed by the right ventricle of a man at rest is approximately 0.2×107 ergs. The stroke work index would therefore be about 0.12×107 ergs/m^2.

178. **(A)** A more than twofold increase in stroke volume either seldom or never develops. Usually the maximum increase in stroke volume achieved is about 50%.

179. **(B)** Tension produced by a contractile sphere (dynes/cm) is determined as follows:

Tension = [pressure produced (dynes/cm^2)] × radius (cm) × 1/2

Before: $(110 \times 1330 \times 4 \times 1/2) = 290{,}000$

After: $(100 \times 1330 \times 5 \times 1/2) = 330{,}000$

% Change $= (3.3 - 2.9) / 2.9 \times 100 = +14\%$

On the basis of these calculations, we can see that after surgery, although the left ventricle is producing less pressure, it is exerting more tension. This demonstrates that a distended ventricle is at a mechanical disadvantage. In other words, distention results in the ventricle having to develop more tension in order to produce the same pressure. Note that in the above calculations, mm Hg is converted to dynes/cm^2 by incorporating the constant 1330 dynes/cm^2/mm Hg in the calculation.

180. **(C)** The tension exerted on a muscle prior to contraction is called its preload. In the case of the left ventricle, it is directly related to the radius or volume of the ventricle and to the pressure in the ventricle.

181. **(A)** The stimulation of cardiac sympathetic neurons causes a decrease in conduction time. Sympathetic stimulation has a positive inotropic, dromotropic, and chronotropic action on the heart. In other words, it increases (1) cardiac force at any given end-diastolic volume (a shift of the Starling curve to the left = a form of homeometric regulation), (2) the velocity of conduction (decreases conduction time), and (3) the frequency of stimulation (increases heart rate).

182. **(A)** Parasympathetic neurons exert little or no direct influence in controlling ventricular conduction or force. The acetylcholine liberated by parasympathetic neurons going to the atria causes the atria to exert less force at each end-diastolic volume (shifts the Starling curve to the right) and decrease conduction speed at the AV node.

183. **(B)** Within limits, as one increases the Ca^{++}, the heart's contractions strengthen at any given end-diastolic volume. In other words, Ca^{++} shifts the Starling curve to the left. For the most part, Ca^{++} and K^+ act antagonistically. K^+ has a negative inotropic action on the heart. KCl injections cause hypopolarization.

184. **(C)** In the healthy adult, the mean ejection pressure produced by the left ventricle is four to five times that produced by the right ventricle. Because the stroke volume produced by both ventricles in the resting subject is equal, the stroke work and stroke work index of the left ventricle will be four to five times that of the right ventricle.

185. **(A)** An increase in the inotropicity of the left ventricle causes a decrease in the pre-ejection period/ejection time as well as a decrease in end-systolic volume. It will increase V_{max}. V_{max} is the velocity of shortening of a muscle when that muscle has no afterload.

186. **(E)** Because stroke work is a function of mean ejection pressure (approximate mean arterial pressure) and stroke volume, large decreases in arterial pressure and stroke volume (cardiac output divided by heart rate) would lead to a big decrease in stroke work. A decrease in inotropic state would be evidenced by decreases in both mean arterial pressure and stroke volume. Vascular resistance is a function of flow and perfusion pressure. While flow decreased somewhat (from 5 L/min to 4 L/min), perfusion pressure decreased far more, leading to a fall in total peripheral vascular resistance.

187. **(B)** Atrial contraction (i) causes a small increase in ventricular volume and is preceded by a period of diastasis (h). Point f marks the beginning of the period of rapid ventricular filling (g). Atrial contraction is followed by isovolumic systole (j) and ventricular ejection (k).

188. **(C)** N represents mitral valve closure after the ventricle has filled with blood (period M–N is the period of ventricular filling) and before it has markedly increased its pressure. M represents mitral valve opening prior to ventricular filling. L represents aortic valve closure after ventricular ejection (period O–L). O represents aortic valve opening prior to ventricular ejection.

189. **(B)** The anterior mitral leaflet is the most mobile of the two. Papillary muscle contraction assists in valve closure by pulling the leaflets toward the midline. An increase in ventricular pressure at the beginning of ventricular systole contributes to valve closure to only a minor extent. The valve leaflets close partially during diastasis and then reopen again during atrial systole.

190. **(C)** The average height of the loop (the mean ejection pressure) decreased during medication. The medication probably lowered the mean ejection pressure by producing vasodilation. The decreased resistance to ventricular outflow represents a decreased afterload, which permits increased muscle fiber shortening and therefore an increased stroke volume. Although the medication decreased the end-systolic volume from 80 to 40 mL, it did not change the end-diastolic volume from 160 mL. Since neither the end-diastolic volume (160 mL) nor the end-diastolic pressure (5 mm Hg) changed, the preload did not change.

191. **(E)** The change that occurred between 1976 and 1979 is that the subject has developed an increase in his end-diastolic volume associated with a decrease in his stroke work and ejection fraction. In other words, there is a shift of the Frank–Starling curve to the right. This is indicative of a deteriorating cardiac function. Catecholamines such as norepinephrine shift the curve to the left and are said to have a positive inotropic action. Heart failure shifts it to the right and therefore has a negative inotropic action. The elevated ejection pressure can be caused by an increased afterload as a result of aortic stenosis, coarctation of the aorta, or arteriolar constriction. An excessive afterload, preload, metabolic load, or contractile impairment are all possible causes of ventricular failure. Characteristic of ventricular failure is an end-diastolic blood volume greater than 200 mL and an end-diastolic pressure greater than 10 mm Hg.

192. **(D)** When the heart is exposed to an increased load, it will get larger (cardiomegaly). This is caused by the laying down of additional intercellular and intracellular elements. The resultant increase in the size of cells is called hypertrophy. It may be associated with an increase in the number of sarcomeres in series or in parallel, or a combination of the two.

193. **(A)** Later in ejection, ventricular pressure peaks and then begins to fall. Ventricular volume is increasing during filling and is not changing during isovolumic ventricular diastole.

194. **(E)** The stroke work increases more than the O_2 consumption under these circumstances. The three major determinants of myocardial oxygen consumption are as stated. The aortic stenosis increases the pressure head between the left ventricle and the aorta during ventricular ejection. Therefore, the mean ejection pressure in the ventricle is elevated.

195. **(C)** The values given are characteristic of the pulmonary artery.

196. **(F)** The values given are characteristic of the left ventricle.

197. **(M)** The veins hold 60 to 70% of the body's blood volume at any one time.

198. **(A)** The aorta and arteries both show pressure dampering (ie, Windkessel) properties, but the aorta shows it to the greatest extent.

199. **(J)** Aortic valve opening marks the lowest pressure seen in the aorta during the cardiac cycle.

200. **(B)** In diastasis, also called the resting phase of the cardiac cycle, pressure is rising slowly and the ECG is isoelectric.

201. **(I)** The dicrotic notch is also called the incisura.

CHAPTER 8

Smooth Muscle

Questions

DIRECTIONS (Questions 202 through 212): Each of the numbered items or incomplete statements in this section is followed by answers or by completions of the statement. Select the ONE lettered answer or completion that is BEST in each case.

202. Which of the following characteristics apply to both skeletal muscle and all smooth muscle?

(A) depolarize in response to an increased Na^+ conductance
(B) produce action potentials that contain an overshoot
(C) are depolarized by acetylcholine
(D) respond to stimulation by producing a spike potential
(E) all of the above are correct

203. Which of the following characteristics are shared by all smooth muscle?

(A) it is controlled by pacemaker cells outside the central nervous system
(B) it exhibits intercellular conduction
(C) it is controlled by sympathetic neurons
(D) it contracts in response to acetylcholine
(E) none of the above are correct

204. Smooth muscle

(A) is not found in the deltoid muscle
(B) does not exhibit a resting transmembrane potential
(C) has a shorter twitch duration than skeletal muscle
(D) has no well-defined motor end plate
(E) lacks pacemaker cells

205. Which of the following statements is correct regarding multiunit smooth muscle?

(A) the muscle fibers of the uterus are an example
(B) the muscle fibers of the iris are an example
(C) A and B
(D) it is characteristically not controlled by intrinsic pacemakers
(E) B and D

206. Which of the following statements is correct regarding the small intestine?

(A) stimulation of parasympathetic neurons increases the frequency of peristaltic waves in the small intestine
(B) decentralization of the small intestine (ie, cutting its sympathetic, parasympathetic, and sensory neurons) causes a disappearance of segmentation
(C) A and B
(D) decentralization of the small intestine causes a disappearance of peristalsis
(E) B and D

207. Which of the following is correct regarding transection of the thoracic spinal cord?

(A) a loss of volitional control over the urinary bladder
(B) a bladder that is always full and constantly dribbling urine out through the urethra
(C) A and B
(D) a bladder that responds to distention by a reflex emptying that is as complete as that found under normal conditions
(E) A and D

208. Atropine is a drug that prevents acetylcholine from acting on smooth muscle, glands, and the heart. Which of the following effects is correctly produced by atropine?

(A) bronchial constriction
(B) diarrhea
(C) A and B
(D) arteriolar vasoconstriction
(E) B and D

209. Smooth muscle, cardiac muscle, and skeletal muscle are said to be viscoelastic. Which of the following definitions BEST characterizes the properties of an elastic material?

(A) exerts tension in response to distention
(B) exerts tension in response to distention that is, within limits, proportional to the speed of distention
(C) exerts tension in response to distention that is, within limits, proportional to the distance distended
(D) exerts tension in response to distention that is, within limits, proportional to the speed of and the degree of distention
(E) exerts a large quantity of tension in response to a small degree of distention

210. Which of the definitions listed in Question 209 best characterizes a viscous or elastic material?

211. Which of the definitions listed in Question 209 explains why muscle is characterized as viscoelastic?

212. Calmodulin is an acidic peptide which

(A) activates an enzyme that causes the contraction of smooth muscle
(B) activates an enzyme that causes the contraction of skeletal muscle
(C) activates an enzyme that causes the relaxation of smooth muscle
(D) activates an enzyme that causes the relaxation of skeletal muscle
(E) causes the relaxation of either smooth or skeletal muscle

Answers and Explanations

202. (A) Some smooth muscles have action potentials that begin at –50 mv and go to only –10 mv. The overshoot is that part of an action potential that goes to a transmembrane potential of greater than +1 mv. Systemic arterioles are hyperpolarized by acetylcholine. Some produce a plateau potential.

203. (E) The intrinsic muscles of the eye (ciliary and iris muscles), as well as many other smooth muscles, are controlled by impulses originating from the central nervous system. The multiunit smooth muscles (intrinsic muscles of the eye, for example) do not exhibit intercellular conduction. The circular muscle of the iris does not receive a sympathetic innervation. The smooth muscle of arterioles relaxes in response to acetylcholine.

204. (D) The transmitter substance released by a branch of a postganglionic neuron is not restricted to a single cell. The arteries and veins of all skeletal muscle contain smooth muscle. The resting transmembrane potential for smooth muscle averages about –50 mv. Smooth muscle has a twitch duration about ten times that of skeletal muscle. This is due in part to the poorly developed sarcoplasmic reticulum in smooth muscle and therefore to an inability to sequester Ca^{++} rapidly. Some smooth muscle contains pacemaker cells.

205. (E) The uterus is an example of visceral (single-unit) smooth muscle. Multiunit smooth muscle, unlike the smooth muscle of the uterus, characteristically does not have intercellular conduction, hence the name "multiunit."

206. (A) Parasympathetic stimulation has a positive chronotropic action on the intestine and a negative chronotropic action on the heart. Segmentation does not require neurons. Peristalsis does not require nervous connections with the central nervous system. It does, unlike segmentation, require the intrinsic nerve plexi of the small intestine. These plexi continue to function after decentralization and differ from the neurons innervating skeletal muscle fibers in this regard.

207. (C) Some paraplegic patients are able to initiate contraction of the urinary bladder by pinching their thighs or scratching the skin in their genital area, but volitional control over micturition, in the usual sense of the term, is lost. Under normal conditions, there is from 0.09 to 2.3 mL of urine left in the bladder at the end of micturition. Unfortunately, this residual volume is much greater in the paraplegic. This creates an important clinical problem. Because urine supports the growth of microorganisms, bladder infection is common among paraplegic patients.

208. (D) Acetylcholine causes arteriolar vasodilation. Atropine, by blocking postganglionic parasympathetic neurons, causes bronchial dilation, as well as constipation.

209. **(C)** The quantity of tension (dynes/cm) that occurs in response to distention (cm) is expressed by the modulus of elasticity, which is related to the slope of the tension–distention curve. Both viscous and elastic materials exert tension during distention. The tension exerted by a perfectly elastic substance is independent of the speed of distention.

210. **(B)** Viscous materials, such as chewing gum, exert a tension when distended that is independent of the length distended.

211. **(D)** It is known from electron-microscopic studies that when striated muscle is distended the thick and thin filaments slide over one another. It is this internal resistance (ie, friction) in striated muscle, and possibly a similar event in smooth muscle, that is responsible for muscle's viscous characteristics. Viscosity (ie, internal resistance) is also an important factor in fluid dynamics (air flow and blood flow, for example). Here, however, we are concerned with a relationship between pressure and flow rather than tension and linear displacement. This relationship for water and other Newtonian fluids can be expressed as follows:

$$\text{Viscosity} = \text{pressure head} \times (\text{tube radius})^4 \times \text{constant} \, / \, \text{flow} \times \text{tube length}$$

These principles are important for understanding such smooth muscle-containing structures as bladders, blood vessels, ducts, tubes, and tracts. Some of the elements that contribute to the viscoelastic characteristics of muscle include the sarcolemma, cell membrane, collagen and elastin fibers, and actomyosin. These are further classified as being either series viscoelastic components or parallel viscoelastic components. The sarcolemma of skeletal muscle, for example, serves an important parallel viscoelastic function, as do the pericardium of the heart and the elastic and collagen fibers of arteries. Actomyosin, on the other hand, is not only an important contractile component of muscle but also an important contributor to the series viscoelastic characteristics of muscle. During rigor, it either prevents distention or ruptures. During relaxation and contraction it resists distention.

212. **(A)** Ca^{++} binding to troponin C in skeletal and cardiac muscle is part of the activation–contraction coupling process. Calmodulin is structurally similar to troponin C, and in smooth muscle serves an activation–contraction coupling function.

PART IV

Circulation

CHAPTER 9

Arteries, Arteriovenous Anastomoses, and Veins

Questions

DIRECTIONS (Questions 213 through 242): Each of the numbered items or incomplete statements in this section is followed by answers or by completions of the statement. Select the ONE lettered answer or completion that is BEST in each case.

213. Eleven hundred millimicrocuries of tritiated water are injected intravenously into a 50-kg man. If you assume that there is complete mixing and distribution throughout the body and that 200 millimicrocuries of the tritiated water were excreted prior to sampling, what approximate plasma concentration of tritiated water would you expect to find?

(A) less than 15 millimicrocuries/L
(B) 30 millimicrocuries/L
(C) 45 millimicrocuries/L
(D) 60 millimicrocuries/L
(E) more than 80 millimicrocuries/L

214. Nine hundred millimicrocuries of albumin labeled with radioactive iodine are injected intravenously into a lean, 70-kg man. If there is complete mixing in the plasma and no excretion of the albumin, what approximate plasma concentration of the labeled albumin do you expect to find 10 minutes after injection?

(A) less than 29 millimicrocuries/L
(B) 60 millimicrocuries/L
(C) 130 millimicrocuries/L
(D) 260 millimicrocuries/L
(E) 500 millimicrocuries/L

215. A 60-kg patient has a hematocrit of 40% and a plasma volume of 3 L. What is his total blood volume?

(A) 4.0 L
(B) 5.0 L
(C) 6.0 L
(D) 7.0 L
(E) greater than 7.5 L

216. The French physician Jean L. M. Poiseuille (1799–1869) defined the factors that regulate the flow of water through a single, rigid cylindrical tube. Which of the following is correct with regard to flow rate as expressed in the Poiseuille equation?

(A) directly proportional to perfusion pressure
(B) inversely proportional to viscosity
(C) A and B are correct
(D) directly proportional to radius to the third power
(E) A and D are correct

217. Which of the following statements is correct?

(A) the formula, resistance = (P1 – P2)/L, is used to determine accurately resistance in the cardiovascular system
(B) in the above formula, (P1 – P2) = pulse pressure
(C) A and B
(D) the formula, resistance = $(8L\eta)/(\pi r^4)$, is used to determine accurately resistance in the cardiovascular system
(E) B and D

218. Under which of the following conditions is an increase in arterial pressure associated with a decrease in peripheral resistance?

(A) when there is turbulence
(B) when there is an increased hematocrit
(C) when there is vasoconstriction
(D) when there is vasodilation
(E) when there is an increased cardiac output

219. The total resistance offered by three resistances—0.2, 0.4, and 0.2—arranged in parallel would be

(A) 0.8 PRU
(B) 0.4 PRU
(C) 0.3 PRU
(D) 0.2 PRU
(E) 0.08 PRU

220. In the horizontal subject, the greatest pressure head (ie, P1 – P2) exists between

(A) the ascending aorta and the anterior tibial artery
(B) the anterior tibial artery and the anterior tibial vein
(C) the anterior tibial vein and the right atrium
(D) the pulmonary artery and the pulmonary vein
(E) the efferent renal arteriole and the renal vein

221. A 70-kg, 6-ft, normal, healthy subject standing quietly erect for 30 seconds has a mean arterial pressure of 100 mm Hg in the ascending aorta and a venous pressure of 2 mm Hg in the superior portion of the inferior vena cava. What is the pressure in the veins of the dorsum of the foot?

(A) less than 0 mm Hg (ie, below atmospheric pressure)
(B) 2 mm Hg
(C) 4 mm Hg
(D) about 20 mm Hg
(E) above 40 mm Hg

222. A soldier stands at attention for 30 seconds. How can his femoral vein pressure best be decreased?

(A) by decreasing the heart rate
(B) by dilating the systemic arterioles
(C) by constricting the systemic arterioles
(D) by having him hold his breath
(E) by having him take one step forward

223. Venous return of blood to the right heart is normally increased by

(A) increased minute ventilation
(B) increased venous tone
(C) increased cardiac sympathetic tone
(D) all of the above
(E) none of the above

224. Which of the following statements is correct concerning the collagen fibers in a healthy, 25-year-old man?

(A) they are intercellular structures most important in preventing the rupture of an artery
(B) they are not found in arteries
(C) they are not intercellular structures
(D) they are not found in veins
(E) B and D

225. A saccular aneurysm in the abdominal aorta is more likely to rupture than that part of the aorta that carries blood to it because

(A) more tension is exerted on the wall of aneurysm
(B) there is less pressure in the aneurysm
(C) there is less turbulence in the aneurysm
(D) there is more tension exerted on and less turbulence in the aneurysm
(E) there is more pressure and less turbulence in the aneurysm

226. A 10-cm strip of artery with no branches was isolated and clamped centrally and peripherally. The pressure in the strip was 90 mm Hg before and 120 mm Hg after the injection of 3 mL of blood. The compliance of this strip of artery in mL/mm Hg was

(A) 0.001
(B) 0.01
(C) 0.1
(D) 10
(E) 1000

227. Which of the following processes occur with aging (ie, between the ages of 45 and 95)?

(A) an increase in concentration of calcium in the arteries
(B) the valves of the veins become incompetent
(C) A and B
(D) an increase in arterial pulse pressure
(E) A, B, and D

228. Referring to Figure 9–1, which of the following is correct?

(A) pulse pressure is increased in C
(B) arteriolar constriction may have occurred in A
(C) A and B
(D) decreased arterial compliance occurred in E
(E) B and D

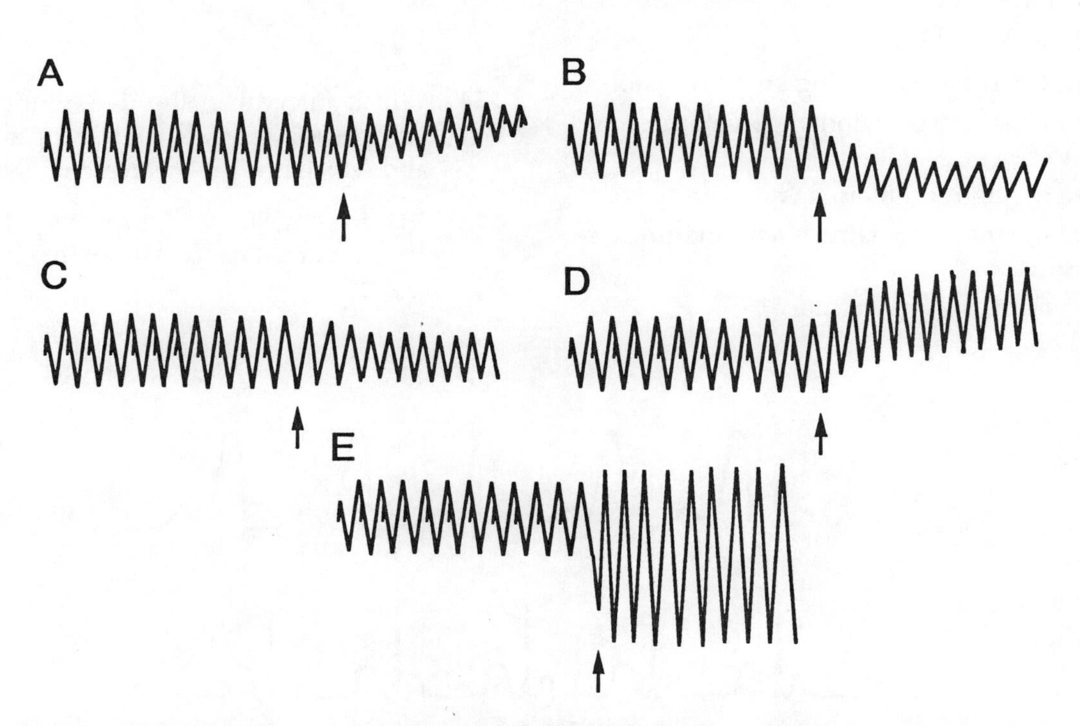

Figure 9–1. Records from a patient with a decentralized heart (all recorded at a constant paper speed and amplification).

229. Aortic insufficiency

(A) increases aortic pulse and diastolic pressure
(B) increases pulse pressure and decreases diastolic pressure
(C) decreases aortic pulse and diastolic pressure
(D) is similar to vasoconstriction in that it decreases pulse pressure and increases diastolic pressure
(E) is similar to vasodilation in that it decreases pulse pressure and increases diastolic pressure

230. An increase in the rate of blood flow from the arterial system causes a more negative slope of the catacrotic limb of the aortic pressure curve. This more negative slope may be caused by

(A) a decreased heart rate
(B) a decreased stroke volume
(C) both of the above
(D) aortic insufficiency
(E) aortic stenosis

231. Referring to Figure 9–2, how would you characterize the pattern?

(A) a bigeminy containing an extra systole
(B) a bigeminy containing a premature systole
(C) a trigeminy containing an extra systole
(D) a trigeminy containing a premature systole
(E) a normal sinus rhythm

232. Which of the following statements about the systemic circulation is correct?

(A) the velocity of flow in the large arteries is from 100 to 500 times faster than in the capillaries
(B) the blood volume in the arteries is approximately equal to that in the veins
(C) A and B
(D) the velocity of flow in the veins is greater than in the arteries
(E) A, B, and D

233. The dye, indocyanine green, is injected rapidly into the cephalic vein of a patient (Fig. 9–3). At the same time, blood samples are withdrawn from the subclavian artery at a constant rate and moved through a cuvette-densitometer. What part of this curve is not an actual recording of dye concentration, but instead represents an estimate (ie, an extrapolation) of the dye-dilution pattern in the absence of recirculation?

(A) line g–h′
(B) curve g–h–i
(C) curve h–i–j
(D) curve i–j
(E) curve i–k

234. Which interval in the dye-dilution curve in Figure 9–3 represents the appearance time and which one the mean recirculation time?

(A) the appearance time is period f–g, and the mean recirculation time is period g–h′

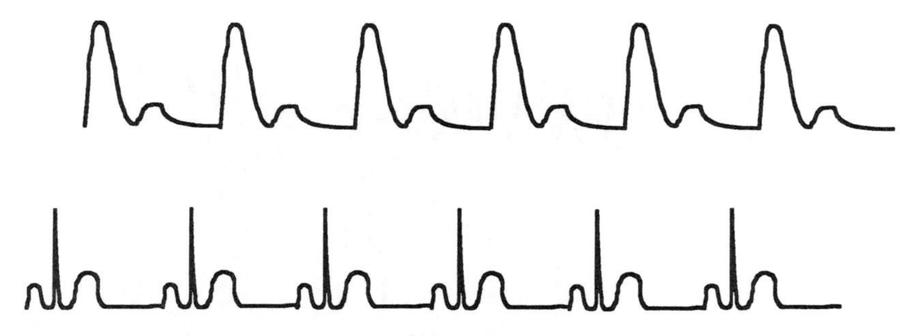

Figure 9–2. Recording from the femoral artery of a patient.

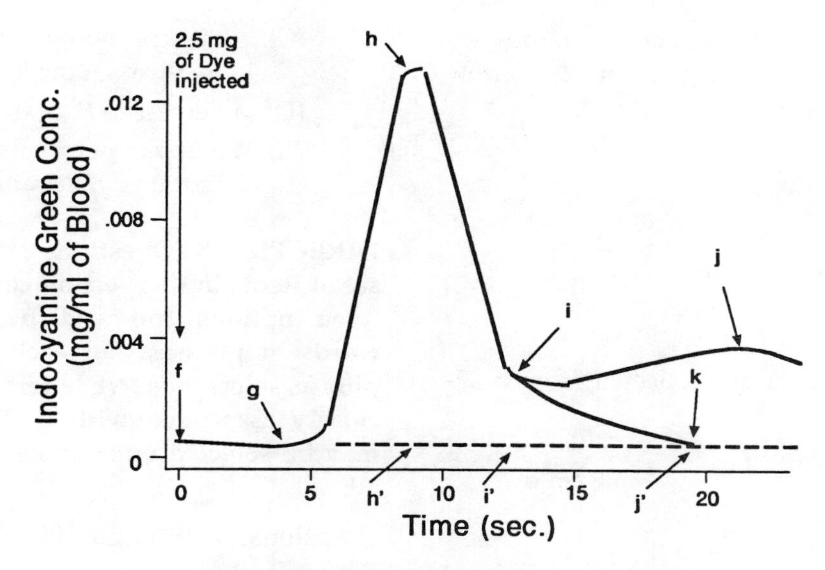

Figure 9–3. Dye dilution curve of a patient.

(B) the appearance time is period f–g, and the mean recirculation time is period h–j′
(C) the appearance time is period f–g, and the mean recirculation time is period i′–j′
(D) the appearance time is period g–h′, and the mean recirculation time is period g–i′
(E) the appearance time is period g–h′, and the mean recirculation time is period h′–j′

235. An investigator can use the dye-dilution curve in Figure 9–3 to estimate the cardiac output of the patient. In doing this, she determines the average concentration under the dye-dilution curve (C = 0.0037 mg/mL), the duration of the curve (T = 15.6 sec), and quantity of dye injected in mg (M = 2.5 mg). What is the cardiac output (mL/min)?

(A) 1560
(B) 1810
(C) 2135
(D) 2500
(E) 2600

236. How would the dye-dilution curve (Fig. 9–3) change in exercise? There would be

(A) an increased mean recirculation time
(B) an increased appearance time
(C) an increased mean recirculation time and appearance time
(D) an increased duration of the total extrapolated curve (ie, period g–k)
(E) none of the above changes

237. How would the dye-dilution curve change in a patient with a femoral arteriovenous fistula (see Fig. 9–3)? There would be

(A) a reduced mean recirculation time
(B) a reduced appearance time
(C) an increased appearance time
(D) no need to extrapolate the recorded curve
(E) a reduced mean recirculation time and an increased appearance time

238. A patient is suspected of having a tetralogy of Fallot that is resulting in a right-to-left shunt of the blood. How would the dye-dilution curve change (see Fig. 9–3)?

(A) a reduced mean recirculation time
(B) a reduced appearance time
(C) an increased appearance time
(D) no need to extrapolate the recorded curve
(E) a reduced mean recirculation time and an increased appearance time

239. Which of the following structures in the resting subject receives the greatest blood flow per gram of tissue?

(A) brain
(B) heart
(C) liver
(D) gastrocnemius muscle
(E) kidney

240. The following data are collected from a 22-year-old patient:

Cardiac output	6 L/min
Total blood volume	5.5 L
Average velocity of flow in systemic arteries	20 cm/sec
Average velocity of flow in systemic capillaries	0.05 cm/sec

What is the total cross-sectional area for the lumina of the patient's systemic capillaries (cm^2)?

(A) 600
(B) 1000
(C) 1300
(D) 1700
(E) 2000

241. Which of the following statements is correct for arteriovenous anastomoses in the skin?

(A) they dilate in response to cutaneous cooling
(B) when dilated, they cause an appreciable reddening of the skin
(C) A and B
(D) when dilated, they cause an appreciable increase in the venous O_2 concentration of associated veins, and they are a means of heat loss
(E) B and D

242. Ligation of the femoral artery in most subjects causes

(A) a cessation of blood flow to the leg
(B) a marked, permanent decrease in blood flow through the leg
(C) a marked, temporary decrease in blood flow through the leg
(D) a decreased P_{CO_2} in the femoral vein
(E) a marked, permanent decrease in femoral vein P_{CO_2} and blood flow

DIRECTIONS (Questions 243 through 248): Each set of items in this section consists of a list of lettered options followed by several numbered words or phrases. For each numbered word or phrase, select the ONE lettered option that is most closely associated with it. Each lettered option may be selected once, more than once, or not at all.

Questions 243 through 245

(A) shear rate
(B) shear stress
(C) kinetic energy
(D) orthostasis
(E) Bernoulli's principle
(F) hematocrit
(G) gravity
(H) potential energy
(I) relative viscosity
(J) Newtonian fluid
(K) PRU
(L) transmural pressure
(M) perfusion pressure
(N) hydrostatics

243. Unitary expression of vascular resistance

244. Force pushing imaginary layers of fluid

245. Ratio of erythrocyte volume to whole blood volume

Questions 246 through 248

(A) viscosity
(B) density
(C) velocity
(D) radius
(E) time
(F) slippage layer
(G) Fahraeus effect

(H) Fahraeus–Lindqvist effect
(I) erythrocyte deformation
(J) Newtonian fluid
(K) maximal oxygen transport
(L) functional turbulence
(M) perfusion pressure

246. Phenomenon whereby viscosity declines in tubes smaller than approximately 1 mm.

247. Phenomenon whereby hematocrit declines in small blood vessels.

248. Condition normal for the aorta due to the high blood flow velocity there, but usually not seen in other large arteries.

Answers and Explanations

213. (B) The average amount of water in the human body is 63% of the body weight. For the approximations required in this problem, you can use values anywhere from 55 to 70% and still select the correct answer.

Concentration in body liquids
= (Quantity injected – quantity lost)/quantity of body water
= (1100 – 200) millimicrocuries/(0.63×50) L
= 29 millimicrocuries/L

214. (D) You can assume that the plasma represents 5% of the body weight and has a density of 1.0 g/mL. You can also assume that, at the time of sampling (ie, 10 minutes after injection), all of the labeled albumin is still in the blood, since the capillaries are relatively impermeable to it. It has been estimated that less than 5% of the radioactive albumin will be lost from the blood after the first hour, less than 60% after the first day.

Plasma volume = 0.05×70 kg $\times$ 1 L/kg = 3.5 L

Plasma concentration = Quantity injected/plasma volume
= 900 millimicrocuries/3.5 L
= 260 millimicrocuries/L

215. (B) The hematocrit is the concentration of red blood cells in the plasma. In this example, the blood is stated to be 40% erythrocytes by volume, and if one assumes that this is an accurate count, then the total blood volume would be:

Blood volume = 100/(100 – Hct) $\times$ plasma volume
= 1.67×3 L = 5.0 L

In the above calculations, the assumption is made that the hematocrit is precisely determined. In most hospitals the stated hematocrit is approximately 5% too high. This is because they make no allowance for the plasma trapped between the erythrocytes. In addition, it has been suggested that the hematocrit in large vessels is higher than that in small vessels. This may result in an additional sampling error that brings the total possible error to 13%. This would mean that the estimated true body hematocrit is 34.8% and the blood volume is 4.6 L.

216. (C) The Poiseuille equation is written as follows:

$$\text{Flow rate} = (P1 - P2) \times (\pi r^4)/8(L)(\eta)$$

where P1 – P2 is the pressure head in dynes/cm^2 (1 mm Hg = 1330 dynes/cm^2), so flow rate is directly proportional to perfusion pressure; r is the radius of the cylinder in cm, so flow rate is directly proportional to radius raised to the fourth power; L is the length of the cylinder in cm, so flow rate is inversely proportional to vessel length; η is the viscosity of the liquid in poises (water has a viscosity of about 10^{-2} poises), so flow rate is inversely proportional to blood viscosity; and π (= 3.1415 . . . .) and 8 are constants.

217. (A) Pulse pressure is systolic minus diastolic arterial pressure. P1 – P2 represents the difference in mean pressures between two points. The formula for resistance in statement (C) is based on a number of assumptions that do not apply to the cardiovascular system. Some of these assumptions are that the system under study consists of (1) nondistensible vessels and (2) cylindrical vessels, and contains (3) a Newtonian fluid. Blood vessels are usually distensible and are conical in shape.

218. (E) A decrease in peripheral resistance tends to decrease arterial blood pressure, but an increase in cardiac output can mask this action:

Arterial pressure (mm Hg)
= [peripheral resistance (PRU)]
× [cardiac output (mL/sec)].

Note: A PRU equals (1 mm Hg)/(mL/sec).

Resistance is defined as the pressure head in mm Hg divided by the flow in ml/sec. In the systemic circulation, the pressure head is approximately equal to the arterial pressure. This ratio (ie, resistance) is increased by turbulence, by an increased hematocrit (ie, an increase in the viscosity of the blood), and by vasoconstriction. In other words, each of these changes increases peripheral resistance. Vasodilation decreases peripheral resistance and, in the absence of a change in flow, also decreases arterial pressure.

219. (E) The total resistance (Rt) for this arrangement is given by the following:

$$1/Rt = 1/Rf + 1/Rg + 1/Rh = 1/0.2 + 1/0.4 + 1/0.2 = 2/0.4 + 1/0.4 + 2/0.4 = 5/0.4$$
$$Rt = 0.4/5 = 0.08$$

220. (B) The small-diameter lumina of this channel (arterioles and capillaries: 7.5 μm) make it a high-resistance system with a pressure loss of about 80 mm Hg. The vessels in the ascending aorta and the anterior tibial artery, though long, have a sufficiently great radius to keep this a low-resistance channel. As a result, the pressure lost in transit is low (ie, about 5 mm Hg). For choice (C), P1 – P2 = 5 to 10 mm Hg; for choice (D), P1 – P2 = 15 mm Hg; and for choice (E), P1 – P2 = 10 mm Hg.

221. (E) Under these circumstances, the pressure in the veins of the foot (PF) is about 91 mm Hg. It will equal the pressure in the vena cava at heart level (PH) plus the change in pressure due to the resistance between the veins of the foot and the veins leading into the heart (PR) plus the pressure due to the weight of the blood (PW). If the distance between the veins of the foot and the heart is 1200 mm, PW equals approximately 88 mm Hg [= (1200 mm H_2O)/(13.6 mm H_2O/mm Hg)]. Therefore PF = PH + PR + PW = 1 + 2 + 88 = 91. The effect of gravity on blood pressure is one of the reasons that edema and overly distended veins (varicose veins) are more common in the lower extremities than the upper extremities.

222. (E) When a subject contracts his skeletal muscles, they tend to compress veins and therefore push blood toward the heart. The valves in the veins of the extremities prevent the blood from flowing away from the heart. In other words, skeletal muscle contraction and relaxation tend to increase the venous return of blood to the heart and decrease the venous pressure. Taking one step forward can decrease the pressure in the veins of the dorsum of the foot by 40 mm Hg.

Most changes in heart rate have little or no effect on central venous pressure. Changes in the caliber of the arterioles will affect how much blood leaves the arterial system and therefore how much enters the venous system. Arteriolar constriction, for example, by increasing arterial blood volume, produces important increases in arterial blood pressure but produces little or no change in venous blood pressure. This apparent discrepancy occurs because the veins are much more compliant than the arteries.

223. **(D)** Venous return is directly related to the amount of push and pull between the great veins, the right atrium, and the right ventricle (ie, [1] the pumping action of the muscles, [2] the compliance of the veins, [3] the pumping action of the heart). When an increase in venous return precedes an increase in cardiac output, the end-diastolic volume of the heart and/or the pulmonary blood volume will increase. Since the circulation is a closed, compliant system, venous return (mL of blood flowing into the heart per minute) can exceed or be less than cardiac output (blood flowing out of the heart per minute) for only short periods (usually less than 1 minute).

224. **(A)** In a young, healthy subject the distensibility of an artery or vein is determined by the elastic fibers and smooth muscles of these vessels. The collagen, under most circumstances, remains slack and does not contribute to distensibility. In cases where an artery or vein is markedly distended, the slack in the collagen is taken up, and it prevents further distention. It performs this function by virtue of its low distensibility and great strength. In this respect, it is similar to the pericardium of the heart. It, of course, differs from the pericardium in that it is a part of the organ it protects from overdistention (ie, the artery or vein), rather than the part surrounding the organ (ie, the heart). Examples of conditions where the collagen fibers have failed to protect from overdistention include saccular and fusiform aneurysms.

225. **(A)** If there were no flow in the system, the pressures would be the same in the lumen of the aneurysm and the afferent vessel. Under these circumstances, according to the law of Laplace, the tension will be greater on the wall of the aneurysm because the radius of the aneurysm is greater than the radius of the afferent vessel:

$$\text{Tension (dynes/cm)} = [\text{pressure (dynes/cm}^2)] \times [\text{radius (cm)}]$$

Since there is flow, the pressure will be greater in the aneurysm than in the afferent vessel.

In a closed system, when the velocity of flow decreases (ie, in the aneurysm), the lateral pressure increases. This inverse relationship between velocity of flow and lateral pressure is called Bernoulli's principle. There is more turbulence in the aneurysm. This can cause a further weakening of the wall and a murmur that can frequently be heard through a stethoscope.

226. **(C)**

$$\begin{aligned}\text{Compliance} &= \text{Change in volume/change in pressure}\\ &= 3\ \text{mL}/30\ \text{mm Hg}\\ &= 0.1\ \text{mL/mm Hg}\end{aligned}$$

227. **(E)** At age 20, about 0.4% of the aorta is calcium, and by age 80 the calcium is in excess of 6%. By age 70, about 70% of the valves of the veins have atrophied. In aging there is a loss of distensibility in the arteries that causes an increase in pulse pressure. In men at age 40, the average arterial pressure is 126/84 (pulse pressure, 42), and at age 80, the arterial pressure is about 147/84 (pulse pressure, 63).

228. **(E)** *Record A.* Arteriolar constriction will impede flow out of the arteries and therefore increase arterial blood volume and aortic diastolic and mean pressures. The decreased pulse pressure is due to the increase in the heart's afterload.

Records A and B. They both represent increases in resistance. In record B this increase causes a decreased aortic diastolic, pulse, and mean pressure, because the stenosis is an increased resistance into, rather than out of, the arterial system.

Record C. Increases in heart rate and stroke volume by increasing cardiac output increase mean aortic pressure and aortic diastolic pressure. Increases in heart rate by increasing the heart's afterload and by decreasing the duration of ventricular filling during each cardiac cycle decrease the stroke volume and aortic pulse pressure.

Record D. The mean aortic pressure is directly related to the volume of blood in the systemic arteries. The aortic pulse pressure is directly related to the change in aortic volume per cardiac cycle. The aortic diastolic pressure is directly related to the volume of blood left in the arteries at the end of diastole. A rise in the stroke volume increases all three parameters.

Record E. A decreased arterial compliance in an areflexic individual represents a decrease in the dampening action of the artery on the ventricular pressure signal, hence the increased systolic pressure. The arteries also lose their ability to store pressure during diastole, hence the decrease in diastolic pressure. If the arteries had 0 compliance, the diastolic pressure would be 0. In other words, the distensibility of the arteries is responsible for much of the energy output of the ventricle being converted to potential energy, which is then changed back to kinetic energy during ventricular diastole. In arteriosclerosis, there is also a decrease in arterial compliance and an increase in aortic pulse pressure, but because other factors are operating, the decreased diastolic pressure usually is not seen.

229. (B) In aortic insufficiency the loss of blood from the arterial system during each diastole is abnormally high, and therefore the aortic diastolic pressure is low and the pulse pressure (= systolic pressure – diastolic pressure) is high. This increased diastolic runoff of pressure during each cycle is also seen when the heart rate, total systemic resistance, or arterial compliance is decreased.

230. (D) In aortic insufficiency the rate of diastolic runoff is increased because the arteries are losing blood to both the capillaries and the relaxing ventricle during diastole. A decreased heart rate causes an increased volume loss per cycle, but not an increased rate of loss. A decreased stroke volume decreases the perfusion pressure across the capillaries and therefore decreases the rate of diastolic runoff of blood and pressure.

231. (E) On the basis of the ECG alone (bottom tracing), one can conclude that this is a sinus rhythm. The pressure pattern is different from what is usually observed in the thoracic aorta but is typical of that found further along the arterial path. As the pressure wave passes down the arterial system, its character is changed by the elastic characteristic of the arterial system. In the horizontal subject, the mean pressure will decrease minutely, the systolic pressure will rise, the diastolic pressure will fall, and the dicrotic wave will separate from the rest of the catacrotic limb of the pressure curve.

232. (A) Under resting conditions, the velocity of flow in the capillaries is about one four-hundredth that in the arteries. This is due to the greater total cross-sectional area of the capillaries (1800 cm^2 vs. 4 cm^2). It means that in the healthy subject, the blood will remain in the capillary long enough to come into equilibrium with the perivascular fluid. The blood volume in the systemic arteries is about 1000 mL, and that in the systemic veins is 3400 mL. The velocity of flow in the large arteries is about 20 cm/sec and in the large veins is about 14 cm/sec.

233. (E) Curve i–k is the extrapolated part of the curve. Line g–h′ is the baseline of dye concentration (ie, optical density) prior to injection. Curve g–h–i is the recorded change in dye concentration. Curve i–j is that part of the recorded curve which is distorted by the recirculation of the dye.

234. (B) The appearance time is the time it takes the tracer (ie, indocyanine green) to move from the point of injection to the sampling densitometer. The mean recirculation time is the interval between the peak of the first dye curve and the peak of the second (the curve due to recirculation of dye past the sampling point).

235. (E) One uses the average concentration under the extrapolated curve to determine cardiac output. In the example shown, the cardiac output would be:

$$\text{Cardiac output} = M/C \times T = 2.5\ \text{mg}/(0.0037\ \text{mg/mL} \times 15.6\ \text{sec}$$
$$= 43.3\ \text{mL of blood/sec}$$
$$= 2600\ \text{mL/min}$$

236. **(E)** In exercise, the velocity of flow increases. Therefore, the mean recirculation time and appearance times will decrease (ie, the dye will move more rapidly through the circulation). The duration of the curve will decrease. In other words, in exercise, there is an increased cardiac output, and therefore the dye will move more rapidly from its injection site through the systemic arteries (ie, its sampling site). Note that, in the formula for cardiac output, a decrease in either C (average concentration under the extrapolated curve) or T (duration of the extrapolated curve) can yield a higher value for cardiac output.

237. **(A)** The systemic circulation can be characterized as having short (coronary circulation), intermediate (circulation through the arm), and long (circulation through the leg) circuits back to the heart. The femoral AV fistula represents a left-to-right shunt in one of the longer circuits. Some of the dye, by taking this shortcut back to the sampling site, causes a shorter mean recirculation time. The appearance time will not change. In this condition, there is little or no alteration in the circulation between the cephalic vein and the sampling site. In this condition, the first and second dye curves will be more fused than ever. Therefore, the need to separate them by extrapolation becomes even more important than in the normal subject.

238. **(B)** There would be a decreased appearance time. In addition, the initial densitometer wave would usually be of smaller amplitude than the second wave.

239. **(E)** The kidney receives 730 mL of blood per g of tissue. The brain receives 60 mL of blood per g of tissue; the heart receives 70 mL of blood per g of tissue; the liver receives 100 mL of blood per g of tissue; and the gastrocnemius muscle receives 5 mL of blood per g of tissue.

240. **(E)** If you assume that all the cardiac output passes through the systemic capillaries, then they receive a flow of 6 L/min (100 cm^3/sec). This is a valid assumption if the arteriovenous anastomoses are closed and there are no abnormal shunts open such as a patent ductus arteriosus. Since these are usually valid assumptions, then the total cross-sectional area of the lumina of patient's systemic capillaries that are in parallel with one another is:

$$\begin{aligned} \text{Cross-sectional area (cm}^2) &= \text{flow (cm}^3/\text{sec)/velocity (cm/sec)} \\ &= 100\ \text{cm}^3/\text{sec}/0.05\ \text{cm/sec} \\ &= 2000\ \text{cm}^2 \end{aligned}$$

241. **(D)** The cutaneous AV shunts do not exchange nutrients with the perivascular space. They function, as far as we know, solely for heat exchange. They also, when open, tend to arterialize the venous blood, but this is not considered one of their functions. Under most circumstances, the body eliminates heat by shunting additional blood to the skin, which warms the skin. If the skin is warmer than the external environment, it will lose heat to that environment. If the external environment is warmer than the skin, the only mechanism for heat loss is through evaporation. AV anastomoses dilate in response to cutaneous warming. These connecting links between arteries and veins, unlike capillaries, are sufficiently thick-walled vessels that one cannot see the red blood in them. One mechanism for warming the skin is an increased nutritive blood flow (ie, an increased flow in the cutaneous blood capillaries). A second mechanism is an increased non-nutritive blood flow (ie, an increased flow through the AV anastomoses).

242. **(C)** After ligation, pre-existing channels expand, so that after an hour, flow is approximately 70% of normal. After several weeks, flow is probably back to normal. These data are based on experiments on dogs but are probably also true of human beings. The speed with which flow returns after ligation will vary from one vascular bed to another. It will depend upon (1) the number and characteristics of the collateral channels, and (2) the ability of the body to form new vessels. There are arteries parallel to the femoral artery that carry blood to the leg. Ligation causes an increase in femoral vein P_{CO_2}.

243. **(K)** The peripheral resistance unit is used by physiologists as a measure of vascular resistance.

244. **(A)** Shear rate divided by shear stress is a definition of viscosity.

245. **(F)** The ratio of erythrocyte volume to whole blood volume is usually obtained by separating the formed elements (mostly red blood cells) from the liquid components.

246. **(H)** It is also known as the sigma effect.

247. **(G)** It provides a major explanation for the Fahraeus–Lindqvist effect, whereby dynamic hematocrit is lower in smaller tubes due to erythrocyte skimming and also the greater velocity of the red blood cells than the plasma in the small tubes.

248. **(L)** Functional turbulence is normal and develops because of the high velocity of blood flow and the large caliber of the tube.

CHAPTER 10

Control of the Circulation

Questions

DIRECTIONS (Questions 249 through 278): Each of the numbered items or incomplete statements in this section is followed by answers or by completions of the statement. Select the ONE lettered answer or completion that is BEST in each case.

249. Which of the following circulations will respond to the stimulation of its adrenergic sympathetic neurons with the most intense decrease (ie, in terms of percent of control) in its blood flow?

(A) cerebral
(B) coronary
(C) cutaneous
(D) pulmonary
(E) skeletal muscle

250. Which of the following is correct regarding changes produced by a decrease in pressure in the carotid sinus from 90 mm Hg to 70 mm Hg?

(A) a decrease in the frequency of impulses moving centrally in the glossopharyngeal nerve
(B) a reflex stimulation of cardiac sympathetic neurons
(C) A and B
(D) a reflex stimulation of cholinergic postganglionic sympathetic neurons
(E) A and D

251. During defecation, urination, and the lifting of heavy loads, the phenomenon of straining occurs involving the Valsalva maneuver. During this maneuver, which of the following events occur?

(A) a decrease in aortic pressure
(B) an increase in aortic diastolic pressure
(C) an increase in intrathoracic and intra-abdominal pressure
(D) an increase in heart rate
(E) all of the above

252. The total systemic peripheral resistance is increased in response to

(A) a decreased blood volume (ie, hemorrhage)
(B) changing from a reclining to a standing position
(C) lifting a heavy load
(D) A and B
(E) A, B, and C

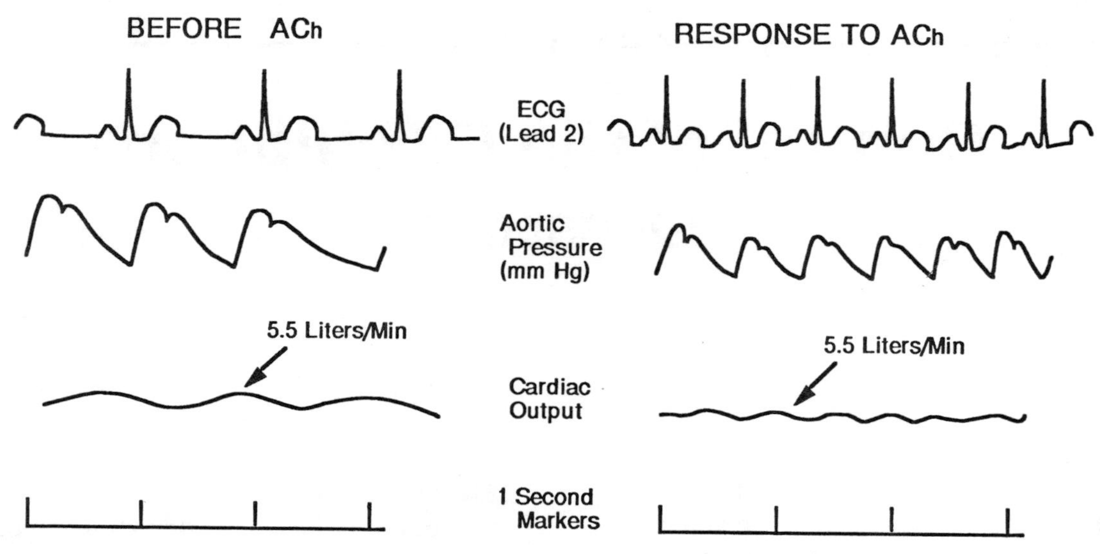

Figure 10–1. A subject's response to acetylcholine (ACh).

253. On the basis of the data presented in Figure 10–1 and an understanding of physiology, one would conclude that acetylcholine produced

(A) an increased pulse pressure
(B) an increased peripheral resistance
(C) a decreased arterial pressure due to venous dilation
(D) an increased systolic pressure
(E) a reflex increase in heart rate

254. You are given three catecholamine solutions of known concentration: norepinephrine, epinephrine, and isoproterenol. Each solution is injected separately into a conscious subject at a concentration of 2 μg/kg of body weight. You can assume that, at this dose, isoproterenol stimulates only β-adrenergic receptors. What response would you obtain from the norepinephrine injection?

(A) an increase in heart rate and arterial pressure that is more marked than that obtained from the other solutions
(B) a decrease in heart rate and an increase in arterial pressure that is more marked than those obtained from the other solutions
(C) decreases in peripheral resistance and heart rate that are more marked than those obtained with any of the other solutions
(D) a decrease in peripheral resistance and an increase in cardiac output that is more marked than with any of the other solutions
(E) an increase in skeletal muscle blood flow and a decrease in cutaneous blood flow

255. In Question 254, what response would be obtained from the isoproterenol injection?

(A) an increase in heart rate and arterial pressure that is more marked than that obtained from the other solutions
(B) a decrease in heart rate and an increase in arterial pressure that is more marked than those obtained from the other solutions
(C) decreases in peripheral resistance and heart rate that are more marked than those obtained with any of the other solutions
(D) a decrease in peripheral resistance and an increase in cardiac output that is more marked than with any of the other solutions
(E) an increase in skeletal muscle blood flow and a decrease in cutaneous blood flow?

256. In Question 254, what response would be obtained from the epinephrine injection?

(A) an increase in heart rate and arterial pressure that is more marked than that obtained from the other solutions
(B) a decrease in heart rate and an increase in arterial pressure that is more marked than those obtained from the other solutions
(C) decreases in peripheral resistance and heart rate that are more marked than those obtained with any of the other solutions
(D) a decrease in peripheral resistance and an increase in cardiac output that is more marked than with any of the other solutions
(E) an increase in skeletal muscle blood flow and a decrease in cutaneous blood flow

257. If a subject puts a rubber band on his arm, pulls it back, and releases it, there occurs the following sequence of events: (1) a red line, (2) a red flare, and (3) a wheal. Which of the following statements is correct?

(A) the red line is due to a multineuron reflex arc
(B) the red flare is due to the release of histamine
(C) A and B
(D) the wheal is due to an axon reflex
(E) none of the above

258. In the heart, brain, skeletal muscle, liver, kidney, and intestine, 1 minute of vascular occlusion is followed by which of the following changes when the occlusion is removed?

(A) a period of 15 to 30 seconds during which flow progressively increases to the preocclusion value
(B) a rapid (less than 2 seconds) return to the preocclusion flow
(C) a period of 15 seconds or more during which the flow is markedly higher than it was prior to occlusion
(D) a change in flow produced primarily by the accumulation of vasodilator metabolites
(E) C and D are correct

259. During early ventricular systole, coronary blood flow

(A) is more markedly reduced at the subepicardial surface than the subendocardial surface
(B) is markedly reduced because the cusps of the aortic valve occlude the coronary arteries and therefore protect them from excess pressure
(C) in the left coronary artery goes to near 0 or below
(D) in the right coronary artery goes to near 0 or below
(E) is well characterized by all of the above statements

260. Which one of the following is most likely to occur in a healthy subject in response to running?

(A) an increased coronary flow due to decreased cardiac adrenergic tone
(B) a generalized increase in blood flow (ie, in the kidneys, muscle, stomach, etc.) due to an increased cardiac output
(C) A and B
(D) a decreased cardiac parasympathetic tone
(E) a decreased velocity of flow in the capillaries of the lungs

261. Which of the following is NOT an important mechanism for increasing the blood flow to active skeletal muscle during exercise?

(A) an increased cardiac output
(B) an increased concentration of epinephrine in the vicinity of the blood vessels of active skeletal muscle
(C) an increased concentration of norepinephrine in the vicinity of the blood vessels of active skeletal muscle
(D) an increased concentration of acetylcholine in the vicinity of the blood vessels of active skeletal muscle
(E) the action of metabolites on skeletal muscle blood vessels

262. Which of the following is most likely to occur in a healthy subject in response to strenuous running?

(A) a decrease in the ejection fraction of the ventricles
(B) a decrease in the maximum dP/dt of the left ventricle
(C) an increase in circulation time
(D) a decrease in arteriovenous O_2 difference
(E) an increase in concentration of arterial lactate

263. In a healthy subject, running is usually associated with a decrease in the end-systolic volume of the right and left ventricles and with an increase in their stroke volume. The mechanism for this response is probably

(A) Starling's law of the heart
(B) an increase in sympathetic tone to the ventricles
(C) an increase in parasympathetic tone to the ventricles
(D) a decrease in venous return of blood to the heart
(E) an increase in pulmonary and systemic resistance

264. A patient with a heart transplant is found to have no reinnervation of his heart. Which of the following responses would you expect him to have to moderate running exercise?

(A) an increased venous return of blood to the heart
(B) an increased heart rate that is more marked than that found in a normal subject during the same exercise
(C) an increased stroke volume that is more marked than that found in a normal subject during the same exercise
(D) an increased stimulation of adrenergic sympathetic neurons to skeletal muscle blood vessels
(E) A and C

265. Pulmonary vascular resistance increases in response to

(A) an elevated P_{O_2} in the pulmonary blood vessels
(B) high altitude
(C) A and B
(D) prostaglandin E_1
(E) none of the above

266. During strenuous running, the cardiac output of a subject increased fourfold, the systemic AV_{O_2} difference changed from 4 to 12 mL/dL, and the mean pulmonary arterial pressure changed from 15 mm Hg to 18 mm Hg. Which of the following conclusions can be drawn?

(A) the increased pressure was caused by an increased pulmonary resistance to flow
(B) the pulmonary artery blood had a lowered P_{O_2} during exercise
(C) A and B
(D) changes in the pulmonary artery blood P_{O_2} are, for the most part, responsible for the changes in the pulmonary peripheral resistance
(E) none of the above

267. Contraction of the smooth muscle of the pulmonary circulation occurs in response to

(A) norepinephrine and epinephrine
(B) serotonin and angiotensin II
(C) A and B
(D) histamine
(E) all of the above

268. The systemic veins differ from the systemic arteries in all of the following ways EXCEPT that the systemic veins

(A) contain a much greater blood volume
(B) are much more compliant
(C) have a more variable blood volume
(D) do not receive a sympathetic innervation
(E) in the arms and thighs contain valves

269. The pulmonary circulation differs from the systemic circulation in that in the pulmonary circulation

(A) the veins do not contain an important reservoir of blood for the heart
(B) the arterial system is more distensible
(C) adrenergic sympathetic neurons are less important in controlling arteriolar and precapillary resistances
(D) the arteries serve as more important blood reservoirs
(E) B, C, and D

270. During running, the cardiac output doubles. What changes would you expect in the pulmonary system?

	Pulmonary Vascular Resistance	Pulmonary Arterial Pressure
(A)	a small decrease	a large increase
(B)	a large decrease	a small increase
(C)	a large decrease	a small decrease
(D)	a small increase	a large increase
(E)	a large increase	a large increase

271. Systemic arterial pressure in the adult is approximately sixfold that of pulmonary arterial pressure because

(A) left ventricular stroke volume is greater than right ventricular stroke volume
(B) systemic blood volume exceeds pulmonary blood volume
(C) systemic resistance exceeds pulmonary resistance
(D) pulmonary compliance exceeds systemic compliance
(E) intra-abdominal pressure exceeds intrathoracic pressure

272. A sustained raised systemic arterial pressure (hypertension) may result from

(A) a decreased secretion of prostaglandin E_1
(B) excessive secretion of adrenocorticotropic hormone (ACTH)
(C) A and B
(D) a decreased peripheral resistance
(E) none of the above

273. Which of the following statements about bradykinin is correct?

(A) it, like histamine, produces vasodilation
(B) it, like angiotensin, is formed from circulating globulins
(C) it is formed by sweat glands, salivary glands, and the pancreas
(D) it, like histamine, produces smooth muscle contraction in the viscera and is capable of attracting leukocytes to an area
(E) all are correct

274. A patient suffers a blood loss over a period of 20 minutes. At the end of this period, his arterial pressure has changed from 100 to 70 mm Hg and his heart rate from 70 to 140/min. His hematocrit is 30% and his skin is cold. What other changes have occurred?

(A) a decreased interstitial fluid volume
(B) an increased total systemic resistance
(C) A and B
(D) an increased plasma colloid osmotic pressure
(E) an increased capillary hydrostatic pressure

275. Other responses to the hemorrhage discussed in Question 274 include

(A) a decreased glomerular filtration rate
(B) a decrease in venous tone
(C) A and B
(D) a decreased secretion of antidiuretic hormone (ADH)
(E) B and D

276. Thirty percent of the blood volume of an anesthetized dog was removed. This was followed by a marked increase in peripheral resistance and a decrease in arterial pressure to 40 mm Hg. Next, the ninth cranial nerve (glossopharyngeus) was severed bilaterally, and the arterial pressure decreased to 20 mm Hg. This second decrease in the arterial pressure was probably due to the destruction of

(A) parasympathetic motor neurons
(B) sympathetic motor neurons
(C) sensory neurons from the carotid sinus
(D) sensory neurons from the carotid bodies
(E) sensory neurons from volume receptors

277. A patient who has lost 30% of his blood volume can be successfully treated if that blood volume is returned to him within the first hour, but he may die if this transfusion is delayed 3 or 4 hours. It has been suggested that this condition is due in part to (1) a baroreceptor-induced reflex vasoconstriction in the viscera, (2) visceral ischemia, (3) deterioration of the capillary barrier in the gastrointestinal (GI) tract, and (4) entrance of bacteria and toxins into the GI capillaries. Which of the following agents would be most effective in relieving the GI ischemia?

(A) epinephrine, because it increases the cardiac output and decreases the peripheral resistance
(B) the related catecholamine, norepinephrine
(C) the β-adrenergic stimulating agent, isoproterenol
(D) acetylcholine
(E) an acetylcholine-blocking agent, such as atropine, for smooth muscle and the heart

278. A patient has suffered a severe hemorrhage at least 4 hours before admission to the hospital. She is given numerous transfusions of blood but is apparently bleeding into the alimentary, respiratory, urinary, and auditory tracts. The laboratory reports that her prothrombin time is infinitely long. How would this bleeding condition best be treated? By administering intravenously

(A) sodium citrate until the prothrombin time is near normal
(B) the antivitamin K drug, dicumarol, until the prothrombin time is near normal
(C) heparin until the prothrombin time is near normal
(D) serotonin
(E) thrombin

DIRECTIONS (Questions 279 through 289): Each set of items in this section consists of a list of lettered options followed by several numbered words or phrases. For each numbered word or phrase, select the ONE lettered option that is most closely associated with it. Each lettered option may be selected once, more than once, or not at all.

Questions 279 through 282

(A) mesenteric circulation
(B) coronary circulation
(C) cutaneous circulation
(D) gastrocnemius muscle circulation
(E) renal circulation
(F) pulmonary circulation
(G) spleen circulation
(H) pancreatic circulation
(I) liver circulation

279. Warming the blood to the hypothalamus of the brain causes a reflex vasodilation in this circulation.

280. A 30% decrease in the total blood volume does NOT cause a reflex vasoconstriction in this part of the systemic circulation.

281. Resistance to flow increases in response to a local decrease in Po_2.

282. Its veins have the highest concentration of oxygen found in the systemic circulation.

Questions 283 through 286

(A) orthostasis
(B) exercise
(C) congestive heart failure
(D) inspiratory tachycardia
(E) hyperventilation
(F) syncope
(G) Valsalva maneuver
(H) Muller maneuver
(I) diving reflex

283. Condition characterized by simultaneous bradycardia and falling arterial blood pressure

284. During stage II, heart rate rises reflexly as arterial blood pressure decreases

285. Condition characterized by simultaneous bradycardia and increasing blood pressure, and elicited by simple breath-hold

286. Produced momentarily and rhythmically by decreased left ventricle stroke volume and arterial blood pressure

Questions 287 through 289

(A) umbilical vein
(B) umbilical artery
(C) ductus venosus
(D) foramen ovale
(E) ductus arteriosus
(F) pulmonary vascular resistance
(G) tetralogy of Fallot
(H) ventricular septal defect
(I) atrial septal defect
(J) pulmonic stenosis
(K) aortic insufficiency
(L) mitral insufficiency
(M) patent ductus arteriosus
(N) aortic stenosis

287. Contains blood of the highest oxygen partial pressure and concentration

288. Congenital condition producing a continuous murmur

289. Congenital condition which is silent

Answers and Explanations

249. (C) The cutaneous and renal vessels have a more pronounced vasoconstrictor response to the stimulation of adrenergic neurons than do the vessels of skeletal muscle and the lungs. The cerebral and coronary circulatory systems are among the least sensitive systems in the body to the constrictor actions of norepinephrine.

250. (C) In the healthy subject, the carotid sinus stretch receptors send progressively fewer impulses up the ninth cranial nerve as the arterial pressure decreases from 200 mm Hg to 30 mm Hg. This results in a progressively greater stimulation of adrenergic sympathetic neurons to the heart and blood vessels and a depression of the cardiac parasympathetic neurons. There is no stimulation of cholinergic postganglionic sympathetic neurons during this reflex.

251. (E) In the Valsalva maneuver, the subject expires against a closed glottis and there results an increased intrathoracic pressure which compresses the arteries and veins, initially increasing arterial pressure. As the increase in thoracic pressure is maintained, venous return is impeded, the cardiac output decreases, and aortic pressure declines. This causes a reflex increase in heart rate and arteriolar constriction, which elevates (increases) aortic diastolic pressure.

252. (E) The reflex response to hypovolemia is increased peripheral resistance. Any maneuver that decreases pressure in the carotid sinus or aortic arch in the healthy, conscious subject (ie, within the range of from 30 to 200 mm Hg) will produce a reflex increase in systemic resistance. This is due to arteriolar vasoconstriction and possibly also to an increased smooth muscle tone in arteries and veins. The Valsalva maneuver and its associated increase in intra-abdominal and intrathoracic pressures occur in response to lifting a heavy load.

253. (E) Acetylcholine has a direct slowing action on the SA node. Because the heart rate increased (ie, there was a decreased R-R interval), apparently, the direct action of ACh is masked by a reflex stimulation of cardiac sympathetic neurons in response to the lowered arterial pressure. The receptors responsible for this reflex are in the carotid sinus and aortic arch. The pulse pressure is reduced. Peripheral resistance = pressure head/cardiac output. Because the pressure head decreased and the cardiac output remained constant, the peripheral resistance decreased. Decreased arterial pressure is usually due to decreased cardiac output or decreased systemic peripheral resistance. Because changes in venous tone do not appreciably affect peripheral resistance, and the cardiac output (and, hence, the venous return of blood to the heart) is constant, the decreased arterial pressure is probably due to arteriolar vasodilation and not to any changes in the venous system. Systolic pressure is peak arterial pressure, which decreases in response to acetylcholine.

254. (B) Norepinephrine produces a more marked increase in the total systemic peripheral resistance than do the other catecholamines. It performs this function by virtue of being a strong stimulator of the

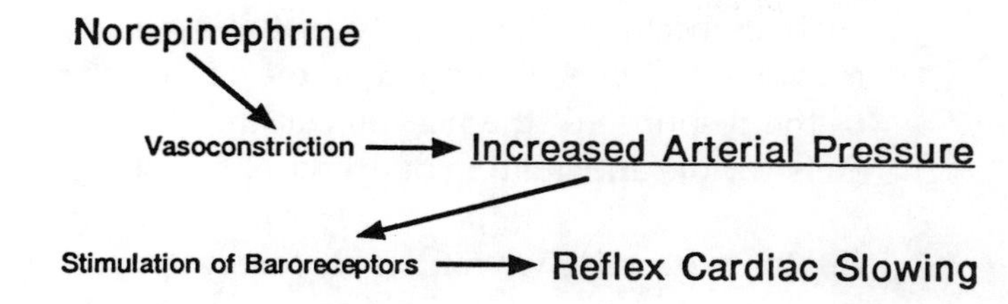

Figure 10–2. Norepinephrine effects on blood pressure.

α receptors of arterioles and, at these concentrations, either not a stimulator or a very weak stimulator of the β receptors of smooth muscle. Although norepinephrine also stimulates the β-adrenergic receptors of the heart to increase heart rate, the increase in arterial pressure that it produces acts through the arterial baroreceptors to mask this action reflexly and produce a slowing of the heart (Fig. 10–2).

255. (D) The response obtained from an isoproterenol injection would be a decrease in peripheral resistance and an increase in cardiac output that is more marked than with any of the other solutions (see Fig. 10–3).

256. (E) Epinephrine has the capacity to produce both vasoconstriction and vasodilation (Fig. 10–4). At the dose injected, the skeletal muscle vasodilation effect would be expected to predominate over the skeletal muscle vasoconstriction effect. The reverse is true in the skin.

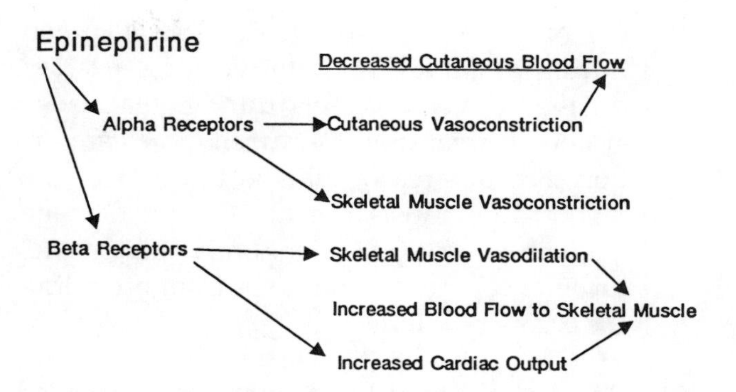

Figure 10–4. Epinephrine effects on blood pressure.

257. (E) The red line is due to a direct response of the vessels to the trauma. The flare is due to an axon reflex. The wheal is due to an intracellular release of histamine, which causes edema.

258. (E) A period during which the flow is markedly higher than it was prior to occlusion is called reactive hyperemia. In the heart, for example, 10 seconds of coronary artery occlusion causes, on release of the occlusion, 15 seconds during which the flow is two to four times that prior to occlusion. Reactive hyperemia occurs in the presence of antihistaminics, in decentralized organs, and in the presence of atropine and adrenergic-blocking agents. Hypoxia, hypercapnea, acidity, and local hyperkalemia may all contribute to the vasodilation found during reactive hyperemia.

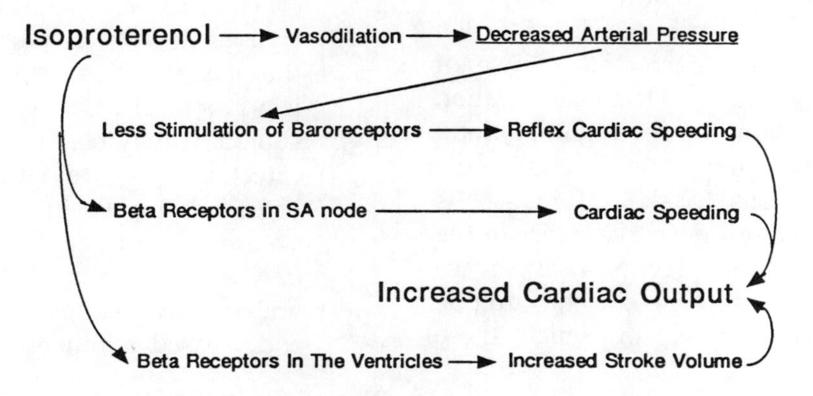

Figure 10–3. Isoproterenol effects on blood pressure.

259. **(C)** Negative flow (back flow) is common during systole. The occlusive force exerted on the coronary vessels during ventricular systole is greatest at the subendocardial portion. For this reason, this is the part of the myocardium most prone to ischemic damage and subsequent infarction. Eddy currents (ie, turbulence) prevent the cusps from occluding the coronary arteries.

260. **(D)** An increased heart rate during exercise is due to a decreased cardiac parasympathetic tone and an increased cardiac sympathetic tone. During exercise, there is an increased cardiac adrenergic tone to the heart. During light exercise, a 300 to 400% increase in skeletal muscle blood flow and no change in renal blood flow may occur. In response to more strenuous exercise, the increased skeletal muscle flow is more marked. In other words, during exercise, at the same time that skeletal and cardiac muscle vessels are dilating, many other vessels are constricting. There is an increased velocity of flow in the pulmonary capillaries.

261. **(C)** The major factors regulating skeletal muscle blood flow during exercise are changes in the distribution of cardiac output initiated by vasodilation in the active skeletal muscle associated with vasoconstriction in most other areas and an increased cardiac output. In strenuous exercise, epinephrine facilitates the shunting of blood to active skeletal muscle by causing vasodilation in skeletal muscle and vasoconstriction in some other areas and an increased cardiac output. In exercise, there is a decreased adrenergic sympathetic tone to the blood vessels of active skeletal muscle. This decreased release of norepinephrine produces a local vasodilation and therefore a shunting of additional blood to this area. The stimulation of cholinergic sympathetic neurons to skeletal muscle is probably an initial response to exercise or the anticipation of exercise. Hypoxia, hypercapnea, and acidity produce skeletal muscle vasodilation. The degree to which these changes occur in the vicinity of the skeletal muscle arteriole will depend on the adequacy of the response of the nervous system to exercise or the anticipation of exercise.

262. **(E)** In an exercise where the O_2 consumption is increased ninefold, the concentration of plasma lactate will double. In running, there is a decrease in the end-systolic volume and an increase in the ejection fraction of the ventricles. Catecholamines are released during exercise and have a positive inotropic action on the heart. The increase in force of cardiac contraction during exercise causes an increase in velocity of blood flow. The AVO_2 difference increases.

263. **(B)** An increase in sympathetic tone to the ventricles shifts the Starling curve to the left. In other words, there is an increase in stroke volume that produces a decrease in the end-systolic volume of the ventricles. Starling's law for the healthy heart states that increases in stroke work (stroke volume × mean ejection pressure) are produced by increases in end-diastolic volume. The Starling mechanism, by itself, does not decrease the end-systolic volume. The parasympathetics play a minor role in the direct control of ventricular stroke volume. They shift the Starling curve for the atria to the right. In running, there is an increase in venous return of blood to the heart and a decrease in pulmonary and systemic resistance.

264. **(E)** There is an increased venous return of blood to the heart as a result of skeletal muscle contraction and relaxation. There is an increased heart rate in response to exercise in the patient with a decentralized heart, but it is a less marked increase than in the normal subject. It may be, in part, due to circulating catecholamines, to which the decentralized heart is hypersensitive. The increased stroke volume is, in large part, due to an increased end-diastolic volume of the heart caused by an increased venous return of blood. There is a decreased adrenergic sympathetic tone to skeletal muscle in both this patient and the normal subject.

265. **(B)** The control over the pulmonary circulation is, in many respects, different from that for the systemic circulation. Thus, hypoxia has opposite effects on systemic arterioles and pulmonary precapillary sphincters. The advantage of such a situation is that poorly ventilated parts of the lung will have some of their blood diverted to the better ventilated parts. If this did not occur, there would be a situation comparable to a right-to-left shunt and therefore a lower saturation of the blood with O_2 in the systemic circulation. Prostaglandin E_1 produces a decreased resistance to flow in both the pulmonary and systemic circulations.

266. **(B)** Using the formula, resistance = pressure/flow, we must conclude that resistance decreased, since flow increased fourfold and pressure 0.2-fold. In other words, the increased flow causes the increased pressure (pressure = resistance × flow). The increased AVo_2 differences during exercise in healthy individuals are due either solely or primarily to a decreased Po_2 in the systemic veins. The decreased Po_2 of the pulmonary artery blood causes precapillary constriction. The decreased resistance to flow during exercise is due to the distention of the vessels by the increased quantities of blood entering them from the right heart and by the high Po_2, characteristic of the alveoli and bronchioles during most exercises.

267. **(E)** Pulmonary arterioles constrict in response to norepinephrine, epinephrine, and angiotensin II. Pulmonary venules constrict in response to serotonin and histamine.

268. **(D)** It has been suggested that increasing adrenergic sympathetic tone to the veins is a mechanism for decreasing venous compliance and shifting blood to the arterial side of the circulation. In a 70-kg man, the systemic veins (3440 mL of blood) contain over three times as much blood as systemic arteries (1000 mL). During hemorrhage and water loading, the volume of the veins changes more than that of the arteries. Because of this, the systemic veins are an important volume buffering system for the arteries. The only arterial valves are the aortic and pulmonary valves. These serve to separate the arterial blood from that in the ventricles.

269. **(E)** Both the pulmonary arteries and veins serve as important reservoirs of blood that the left heart can call upon. The blood volume in the pulmonary system of an adult may vary from 200 mL to 1000 mL. In the pulmonary circulation, unlike the systemic circulation, the capillaries represent the major source of resistance, and many of the arterioles are almost devoid of smooth muscle. Apparently, the main effect elicited by the stimulation of adrenergic sympathetic neurons to the pulmonary blood vessels is to decrease the compliance of arteries and veins (ie, to stimulate their smooth muscles). Almost as much blood volume lies in the pulmonary arteries as in the pulmonary veins.

270. **(B)** A doubling of the cardiac output is usually associated with a 2- to 5-mm Hg increase in pulmonary artery pressure. This represents approximately a 40% reduction in pulmonary resistance. The large compliance of the pulmonary circuit results in large decreases in resistance in response to increases in right heart output. In the systemic circuit, on the other hand, adrenergic sympathetic tone to the arterioles is a much more important factor in the control of resistance.

271. **(C)** Pressure = resistance × flow. Since the flow from the left ventricle approximately equals the flow from the right ventricle, resistance becomes the determining factor. Right and left ventricular stroke volumes are approximately equal.

272. **(C)** Prostaglandin E_1 is a vasodilator. It has been suggested that the hypertension that occurs after removal of both kidneys (renoprival hypertension) is due to decreased levels of circulating prostaglandins. ACTH facilitates the release of cortisol, which increases the sensitivity of the body to norepinephrine and epinephrine. Cortisol may also cause Na^+ and water retention by the body. In Cushing's syndrome, plasma cortisol values are elevated, and about 85% of the individuals with this malady are hypertensive. An increased peripheral resistance tends to increase arterial pressure.

273. **(E)** All of the statements are correct.

274. **(C)** A decreased interstitial fluid volume and an increased total systemic resistance have also occurred (see Fig. 10–5).

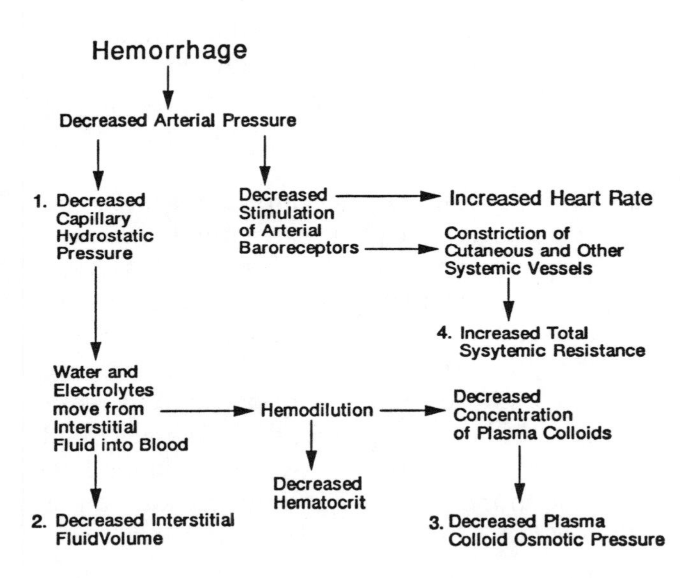

Figure 10–5. Effect of hemorrhage on cardiovascular functions.

275. **(A)** Another response to hemorrhage is a decreased glomerular filtration rate (see Fig. 10–6).

276. **(D)** Under the conditions of the experiment, there should be an intense stimulation of these hypoxia-sensing receptors that will not only affect the medullary respiratory centers but will also spill over into its neighbor, the vasomotor center. When this stimulation to the vasomotor center is lost, arterial pressure

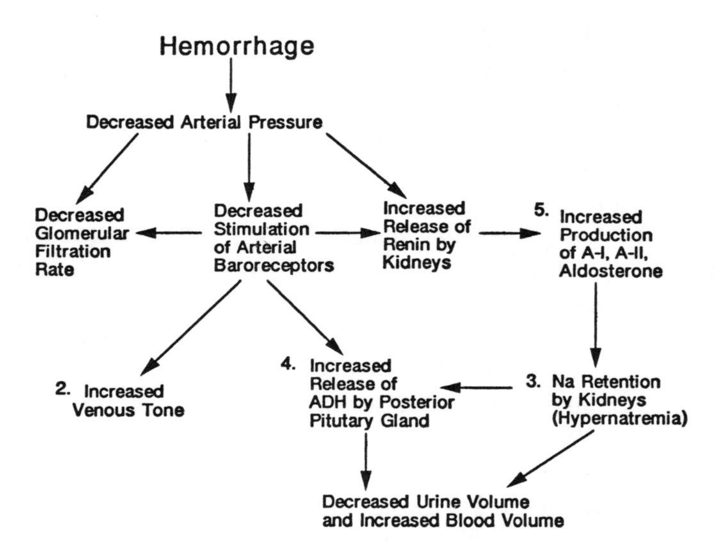

Figure 10–6. Response to hemorrhage.

will decrease. Under resting conditions, in contrast, the carotid bodies play little or no role in the control of blood pressure. Parasympathetic motor neurons in the glossopharyngeal nerve play essentially no role in the control of arterial pressure. There are no sympathetic neurons in the ninth cranial nerve. At pressures as low as 40 mm Hg, the aortic pressoreceptor message level to the brain is near 0. At higher arterial pressure levels, decentralization of the carotid sinus causes an elevated arterial pressure.

277. **(C)** It will, by increasing cardiac output, tend to maintain arterial pressure while decreasing resistance to flow in the GI tract. Unfortunately, like all catecholamines, it may produce dangerous arrhythmias of the heart. Epinephrine and norepinephrine cause vasoconstriction in the GI tract and therefore may not relieve its ischemia. Acetylcholine will lower peripheral resistance to such an extent that there is no pressure left to maintain what little blood flow is already present. It, unlike isoproterenol, does not elevate cardiac output. Acetylcholine is not the cause of the problem in hypovolemia.

278. **(C)** Heparin is an anticoagulant that interferes with the formation of thrombin. In this case, it would be used to prevent prothrombin and fibrinogen consumption and would be withdrawn when the prothrombin concentration was at a more effective level. Sodium citrate would produce a hypocalcemia and a resultant hyperreflexia that might be fatal. It is most often used as an anticoagulant for the storage of blood outside the body. One of the reasons for this problem may be a disseminated intravascular coagulation, which depletes the body of prothrombin as fast as it is added to the blood. Dicumarol interferes with prothrombin production by the liver. The vasoconstriction produced by serotonin would be counterproductive. Thrombin would cause the production of clots at or near the point of injection rather than at the points where bleeding is occurring.

279. (C) The heat regulatory center in the hypothalamus increases heat loss from the body by decreasing adrenergic sympathetic tone to the blood vessels of the skin.

280. (B) Autoregulation is a major factor in the control of both cerebral and coronary blood flow. For example, a local elevation of P_{CO_2} acts as an important cerebral vessel dilator. Vasomotor reflexes play little or no role in the regulation of either cerebral or coronary blood flow, but increases in metabolism or decreases in blood flow do cause an accumulation of metabolites and a resulting vasodilation.

281. (F) In the pulmonary circulation, a low alveolar P_{O_2} or pH causes a local vasoconstriction. This serves to shunt blood away from poorly ventilated alveoli.

282. (E) The high blood flow to the kidneys is not used to meet their metabolic needs (hence the high oxygen concentration in the renal vein), but rather to permit the removal from the blood of substantial quantities of wastes and excess electrolytes and water.

283. (F) The so-called vasovagal faint may be precipitated by stressful or painful situations, such as venipuncture, surgical manipulations, or sight of blood. It is exacerbated by hunger, fatigue, and heat. Fainting may also be caused by orthostatic hypotension, Valsalva maneuver, hyperventilation, reflex cardiac standstill, ventricular fibrillation or heart block, carotid sinus massage, cerebral arterial occlusion, etc. In some individuals fainting takes place following micturition (urination), defecation, deglutition (swallowing), coughing, eyeball pressure, head turning, or with slight physical effort.

284. (G) The Valsalva maneuver, named for the early eighteenth-century anatomist, Antonio M. Valsalva, involves forcing against a closed glottis or mouth. In stage II, cardiac venous return is decreased, reducing ventricular preload, hence stroke volume and blood pressure, which reflexly raises heart rate. This phase may continue for many seconds, with both the hypotension and the tachycardia becoming progressively more extreme. Hypotension and tachycardia are present at this time despite the fact that the Valsalva maneuver also involves breath-holding, which usually produces opposite changes in blood pressure and heart rate.

285. (I) The diving response is largely elicited by apnea (breath-holding) alone. It is further intensified by face immersion, or submersion of the whole body in water. Apnea is directed by higher central nervous system (CNS) centers, which act on the medullary respiratory center. Under its influence and the peripheral chemoreceptors (carotid and aortic bodies), the vasomotor center increases peripheral sympathetic tone, which causes systemic arteriolar vasoconstriction and raises blood pressure. This transient modest hypotension acts on the baroreceptors, producing increased parasympathetic tone to the heart, resulting in bradycardia. The vasomotor center also acts directly on the cardioinhibitory center to induce cardiac slowing.

286. (D) Inspiratory tachycardia is a normal phenomenon which goes on constantly in resting subjects. As the name indicates, heart rate rises during inspiration, while blood pressure falls. During inspiration, intrapleural pressure decreases, increasing right heart venous return and preload. This increases right ventricle stroke volume through the Frank–Starling mechanism. Blood reaching the lungs, however, tends to stay in the lungs as the air and blood capacity is increasing. This decreases venous return, hence preload in the left ventricle. Increased right ventricle preload also moves the interventricular septum to the left, limiting left ventricle filling capacity. The net result is a decrease in left ventricle stroke volume and therefore a momentary fall in arterial blood pressure. A rapid reflex increase in heart rate via the medullary cardiovascular centers ensues, as parasympathetic inhibition of the heart is relieved and sympathetic drive is increased.

287. **(A)** The umbilical vein contains blood which has just been oxygenated by passage through the fetal side of the placenta.

288. **(M)** Blood continually flows through the patent ductus arteriosus from the aorta to the pulmonary artery, producing a continuous murmur (sound).

289. **(I)** Blood flow through an atrial septal defect (ASD) occurs at such low pressure that little sound is produced.

CHAPTER 11

Capillaries and Sinusoids

Questions

DIRECTIONS (Questions 290 through 307): Each of the numbered items or incomplete statements in this section is followed by answers or by completions of the statement. Select the ONE lettered answer or completion that is BEST in each case.

290. What structural feature do all blood capillaries have in common?

(A) absence of intracellular fenestrations in the endothelial cells
(B) presence of intracellular fenestrations
(C) a discontinuous endothelium
(D) a continuous basement membrane
(E) they remain patent in the healthy subject

291. Which structures in the body are normally devoid of blood capillaries?

(A) cartilaginous plates of long bones
(B) lenses of the eyes
(C) the cartilaginous plates and the lens
(D) corneas
(E) dermis of the skin

292. Which of the following best characterizes lymph capillaries?

(A) they have a smaller diameter than blood capillaries
(B) they are less permeable than blood capillaries
(C) they have no endothelial lining
(D) they have a discontinuous basement membrane
(E) they are found only in the liver, spleen, bone marrow, lymph nodes, and gastrointestinal tract

293. Which of the following best characterizes the sinusoids?

(A) they have a smaller diameter than lymph capillaries
(B) they are not found in skeletal muscle
(C) they have a continuous endothelial lining
(D) they have a continuous basement membrane
(E) they are less permeable than lymph capillaries

294. Carbon dioxide moves readily from the extravascular space into the blood. The mechanism or mechanisms responsible for the permeability of skeletal muscle capillaries to CO_2 is

(A) the lipid solubility of CO_2
(B) pinocytosis
(C) diapedesis
(D) endothelial fenestrae
(E) active transport

295. If radioactive urea were injected into the right femoral artery, the time it would take for it to reach equilibrium in the various interstitial fluids would vary. The interstitial fluid into which it would move most slowly would be that found in the

(A) brain
(B) heart
(C) left leg
(D) right leg
(E) kidney

296. A substance was injected intravenously and found to be distributed through 30% of the body water. It probably

(A) did not pass freely through the blood capillaries
(B) was distributed uniformly throughout the body water
(C) did not enter the cells of the body
(D) was not excreted or utilized in the body
(E) was excluded from the cerebrospinal fluid

297. The following data were collected from a skeletal muscle capillary:

Capillary hydrostatic pressure	30 mm Hg
Tissue hydrostatic pressure	2 mm Hg
Effective osmotic pressure	23 mm Hg
Capillary colloid osmotic pressure	25 mm Hg
Tissue colloid osmotic pressure	2 mm Hg

The filtration pressure in this capillary was

(A) greater than 23 mm Hg pushing fluid out of the capillary
(B) between 17 and 23 mm Hg pushing fluid out of the capillary
(C) between 9 and 16 mm Hg pushing fluid out of the capillary
(D) between 1 and 8 mm Hg pushing fluid out of the capillary
(E) negative, that is, there was a net influx into the capillary

298. A patient exhibits swelling of the ankles and a bloated abdomen and has a history of malnutrition. The bloated abdomen is probably due to

(A) increased intestinal gas
(B) slow, chronic abdominal hemorrhage
(C) increased capillary hydrostatic pressure
(D) increased capillary colloid osmotic pressure
(E) decreased capillary colloid osmotic pressure

299. Blood flow (Q), venous oxygen concentration (CvO_2), and interstitial fluid volume (ISF) were measured in the leg. The metabolism of the structures in the leg was kept constant as was the arterial Po_2 and pressure. Under these circumstances, either an increased venous pressure or a decreased vasoconstrictor tone to the leg would cause

(A) an increased Q
(B) an increased CvO_2
(C) an increased ISF
(D) an increased Q and CvO_2
(E) an increased Q and ISF

300. The loss of fluid by the systemic blood capillaries in a healthy subject is

(A) more than the gain of fluid by the capillaries
(B) usually increased by a local arteriolar dilation
(C) A and B
(D) decreased by the stimulation of adrenergic sympathetic neurons to the veins
(E) increased by the dilation of postcapillary sphincters

301. Which of the following is least likely to cause edema in the adult?

(A) nephritic syndrome
(B) congestive heart failure
(C) liver malfunction due to chronic alcoholism

(D) mastectomy
(E) administration of thyroxine to a hypothyroid patient

302. Which of the following is (are) correct regarding the mechanisms responsible for producing edema in the following conditions?

(A) nephritic syndrome—there is an increased venous pressure that causes an increased capillary hydrostatic pressure
(B) congestive heart failure—a loss of albumin in the urine causes a reduction of the capillary colloid osmotic pressure
(C) liver malfunction due to chronic alcoholism—there is an increased venous pressure that causes an increased capillary hydrostatic pressure
(D) mastectomy—the lymphatic drainage from the arms is markedly decreased
(E) A, B, and C

303. The blood capillaries and systemic veins are more important than the lymph capillaries and ducts in the transportation of

(A) cholesterol and stearic acid from the intestine
(B) glucose from the intestine
(C) cholesterol, stearic acid, and glucose from the intestine
(D) albumin from the intercellular space of the intestine
(E) renin from its site of production in the kidney

304. The lymph ducts differ from the veins in that the lymph ducts

(A) are devoid of semilunar valves
(B) contain a fluid with a higher velocity of flow
(C) contain a fluid almost devoid of glucose
(D) contain a fluid devoid of leukocytes
(E) have none of the above characteristics

305. In a healthy individual, two substances, f and g, are found to have a lower concentration on the venular side of a capillary than the arteriolar side. Their concentration in the blood changes as they pass through the capillary (Fig. 11–1). What valid conclusions can you make from this figure?

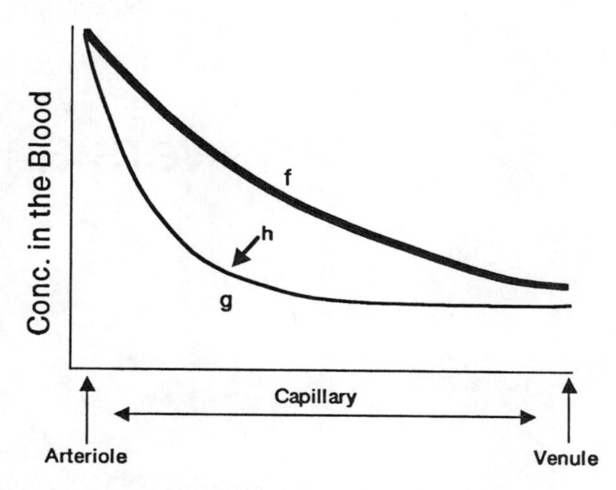

Figure 11–1. Change in concentration of two substances, f and g, as they pass through a capillary bed.

(A) substance f is blood flow-limited (ie, perfusion-limited)
(B) substance g is diffusion-limited
(C) A and B
(D) substance g could be O_2
(E) all of the above

306. A local constriction of the cutaneous arterioles of the hand can cause a local

(A) increase in cutaneous capillary flow
(B) increase in capillary transmural pressure
(C) A and B
(D) complete closure of the capillary lumen
(E) increase in lymph flow

307. As blood passes through the capillary beds of skeletal muscle, there is

(A) a net flux of Cl^- out of the red blood cell (RBC)
(B) an increased affinity of hemoglobin for O_2
(C) A and B
(D) an increased affinity of hemoglobin for H^+
(E) A and D

Answers and Explanations

290. (D) A continuous basement membrane helps to prevent overdistention of blood capillaries. The blood capillaries of the heart, skeletal muscle, lung, nervous system, dermis, subcutaneous and adipose tissue, and the placenta are devoid of a fenestrated endothelium. Fenestrations are found in the capillaries of the glands, renal glomerulus, and vagina. The cerebral capillaries apparently have a continuous endothelium with no apparent intercellular space between one endothelial cell and the next. At any one point in time, most of the blood capillaries in the body are collapsed. It has been noted that during strenuous exercise, the number of patent capillaries in an active muscle may be five times that in the resting state.

291. (C) The cartilaginous plates and the lens are normally devoid of blood capillaries. They are sparsely concentrated in the cornea.

292. (D) The diameter of patent blood capillaries is about 7 μm and that for lymph capillaries may exceed 14 μm. Albumin passes much more freely through lymph capillaries. They are found throughout the body adjacent to the blood capillaries (ie, in skeletal muscle, the heart, the dermis of the skin, etc.).

293. (B) Most of the sinusoids of the body are in the liver, bone marrow, and spleen. Some of the spaces between the cells that line their lumen exceed 1 μm. The sinusoids of the bone marrow appear to lack any basement membrane. The sinusoids of the spleen have a discontinuous basement membrane.

294. (A) Carbon dioxide moves through the capillary barrier more readily than do gases of comparable or smaller size (O_2, for example). Particles with a radius of less than 0.03 μm can be transported across the endothelial lining by microphagocytosis. This is a process that is unimportant for the transport of gases, hexoses, and water, because it is incapable of moving large quantities of any molecule. It is important in moving large molecules to which the capillary is impermeable and which are important in trace amounts (vitamin B_{12}, for example). Diapedesis is a mechanism by which neutrophils and other wandering cells pass the endothelial barrier. The endothelium of the capillaries of skeletal muscle is devoid of fenestrae.

295. (A) With the notable exception of CO_2, most substances do not move readily between the blood and the intercellular fluid of the brain. There is said to be a blood–brain barrier. Part of the reason for this barrier is the processes of the astrocytes that surround the cerebral capillaries. Urea in the blood takes more than 5 hours to come into equilibrium with the intercellular fluid of the brain. In the case of the other tissues mentioned, this occurs in less than an hour.

296. (C) Because approximately 7% of the body water is plasma and 21% is interstitial fluid, it seems reasonable to assume that if a substance is restricted to 30% of the body water, it is not entering the body cells to an important degree, since the body's cells contain approximately 72% of the body water.

297. (D) If you let a plus (+) represent a pressure moving fluid out of and a minus (–) represent a pressure moving fluid into a capillary, then:

Effective hydrostatic pressure = 30 – 2 = +28 mm Hg

Effective osmotic pressure = 2 – 25 = –23 mm Hg

Filtration pressure = 28 – 23 = +5 mm Hg

298. (E) Vitamin deficiency, amino acid deficiency, caloric deficiency, liver damage, or excessive loss of plasma proteins in the urine can produce a decreased capillary colloid osmotic pressure, which causes an increased movement of water from the plasma and into the abdomen and other intercellular spaces. The other possibilities are either incorrect or highly unlikely.

299. (C) Blood flow through the capillaries is directly related to the perfusion pressure (arterial pressure/venous pressure) and inversely related to the resistance (ie, the degree of vasoconstriction). Increases in capillary hydrostatic pressure produce increases in ISF and lymph flow and are caused by increases in venous pressure or arteriolar dilation.

	Q	CvO_2	ISF
Increased venous pressure	decrease	decrease	increase
Decreased vasoconstrictor tone	increase	increase	increase

300. (C) The lymphatic system returns the liquid lost from the capillaries back to the circulation. Arteriolar dilation permits more of the arterial pressure to be transferred to the capillaries. The importance of adrenergic sympathetic neurons to veins is not fully understood. We do know, however, that under experimental conditions, their stimulation can increase smooth muscle tone, decrease venous compliance, and increase venous pressure. Increases in venous pressure increase the capillary hydrostatic pressure. The dilation of postcapillary sphincters decreases capillary hydrostatic pressure.

301. (E) When hypothyroidism develops in the adult, it is called myxedema (Greek *myxa,* mucus, + *oidema,* swelling). In this condition, there is a high incidence of edema in the (1) face (79%), (2) eyelids (90%), and the (3) periphery (55%). The edema of the skin is, for the most part, due to the accumulation of a variety of protein–carbohydrate complexes (ie, colloids) in the intercellular space and can be relieved by the administration of thyroxine.

302. (D) In the nephritic syndrome, a loss of albumin in the urine causes a reduction of the capillary colloid osmotic pressure. In congestive heart failure, there is an increased venous pressure that causes an increased capillary hydrostatic pressure. In liver malfunction, there may be a decreased production of albumin and therefore a decreased capillary colloid osmotic pressure.

303. (B) Glucose and most other nutrients move readily into the blood capillaries of the intestine. Most of the long-chain fats enter the circulation via the lymphatic system rather than the hepatic portal system. The lymphatic system moves a quantity of albumin approximately equivalent to half of the total plasma albumin into the circulatory system each day. This consists of albumin that has escaped from the blood through the blood capillaries. It moves more readily from the perivascular space into the lymph capillaries than into the blood capillaries because of the greater permeability of the lymph capillaries.

304. (A) The semilunar valves in the veins and lymph ducts facilitate the transport of blood and lymph. It has been estimated that in the adult the thoracic duct moves approximately 4 liters of lymph into the blood per day and the inferior vena cava moves over 7000 liters per day. Thoracic duct lymph in the dog averages 124 mg of glucose/100 mL. In the rat, the thoracic duct carries approximately 109 cells (ie, erythrocytes and leukocytes) to the blood per day.

305. (A) Substances in the blood (f in this case) that do not reach equilibrium with the tissues in their passage through the capillary are diffusion-limited. Substance g has reached equilibrium at point h. Oxygen has a concentration curve similar to g and is flow-limited. The delivery of substance g to the tissues can be increased by increasing blood flow to those tissues.

306. (D) Constriction of precapillary sphincters, metarterioles, and arterioles lying between arteries and capillaries may so lower the pressure in the capillaries to cause the loss of their lumen. The intraluminal, hydrostatic pressure at which vessels collapse is called their critical closing pressure and may vary from 10 to 60 mm Hg in the skin.

307. (D) Deoxyhemoglobin has a greater affinity for H^+ than does oxyhemoglobin. There is a net flux of HCO_3^- out of and of Cl^- into the RBC. There is a decreased affinity of hemoglobin for O_2 owing to the increased H^+ and CO_2 concentration.

CHAPTER 12

Extravascular Fluid Systems

Questions

DIRECTIONS (Questions 308 through 318): Each of the numbered items or incomplete statements in this section is followed by answers or by completions of the statement. Select the ONE lettered answer or completion that is BEST in each case.

308. Which of the following statements concerning cerebrospinal fluid (CSF) is correct?

(A) CSF reduces the weight of the central nervous system
(B) over 30% of the cerebrospinal fluid is formed by the choroid plexi
(C) normally, occlusion of the jugular veins causes a prompt (less than 12 seconds) rise in CSF pressure in the subarachnoid space at the level of the fourth lumbar vertebra
(D) The arachnoid villi that project into the superior sagittal sinus tend to facilitate the movement of fluid into this venous sinus but also tend to prevent filtration from the venous sinus into the subarachnoid space
(E) all are correct

309. Which of the following statements about cerebrospinal fluid is correct?

(A) when CSF pressure increases by more than 5 mm Hg, its formation ceases
(B) CSF tends to produce an exaggerated transmural pressure in the blood vessels of the brain
(C) CSF is a more poorly buffered solution than blood
(D) CSF has approximately the same concentration of Ca^{++} and K^{+} as plasma
(E) all are correct

310. Obstruction of the cerebral aqueduct causes

(A) no important change in CSF pressure in the lateral ventricles of the brain because there are alternate pathways between the first, second, and third ventricles and the fourth ventricle
(B) an important decrease in the CSF pressure in the lateral ventricles
(C) a cessation of CSF production in the lateral ventricles
(D) B and C
(E) an important increase in CSF pressure in the lateral ventricles

311. Intraocular fluid (aqueous humor) differs from cerebrospinal fluid in that intraocular fluid

(A) has a protein concentration similar to that in plasma
(B) does not cause a serious increase in pressure when reabsorption is decreased
(C) is an ultrafiltrate of plasma
(D) serves no important nutritive function
(E) contains a higher concentration of ascorbic acid

312. Which of the following statements is correct regarding intraocular fluid?

(A) it is produced at the canal of Schlemm
(B) it is reabsorbed by a structure in the ciliary process that is similar in structure and function to the choroid plexus
(C) it is well characterized by both of the above
(D) it helps maintain the curvature of the cornea
(E) it is well characterized by all of the above

313. Which of the following procedures is most likely to increase the intraocular pressure of the glaucoma patient and therefore exacerbate the condition?

(A) parasympathetic blockade with atropine
(B) a decreased pressure in the jugular vein
(C) a diet high in vitamin C
(D) an increased intensity of light
(E) carbonic anhydrase inhibitors

314. Where in the body do we find endolymph?

(A) lung lymph capillaries
(B) inside the eye
(C) inside cartilage
(D) cochlear duct
(E) nowhere

315. Where in the body is perilymph found?

(A) inside the eye
(B) scala tympani of the cochlea
(C) lung lymph capillaries
(D) inside cartilage
(E) nowhere

316. Which of the fluids listed below normally have protein concentrations of less than 3 mg/mL?

(A) cerebrospinal fluid
(B) intraocular fluid
(C) plasma
(D) amniotic fluid
(E) A, B, and D

317. Which of the fluids listed below has a K^+ concentration in excess of 140 mEq/L?

(A) cerebrospinal fluid
(B) intraocular fluid
(C) endolymph
(D) amniotic fluid
(E) urine

318. All of the following are important functions served by the amniotic fluid EXCEPT

(A) it reduces the weight of the fetus (buoyancy action)
(B) it serves a major function in respiratory gas exchange
(C) it transports wastes and nutrients in superficial areas; at term, about 600 mL of amniotic fluid is formed per hour
(D) it stores hormones; estrogens are at a concentration of 63 μg/100 mL in amniotic fluid and 15 μg/100 mL in maternal plasma
(E) it protects against trauma (shock absorber action)

Answers and Explanations

308. (E) The brain and cord weigh about 1500 g outside the body but inside the body are buoyed up so that their weight is reduced to 50 g. The major sites for the formation of CSF are probably the choroid plexi in the ventricles of the brain, the glial elements in the ependymal wall of the brain, and the cerebral blood vessels. In the adult, the subarachnoid space extends to approximately S2. Jugular occlusion should cause an increased CSF pressure throughout the subarachnoid space and ventricles of the brain, because it decreases absorption of CSF at the dural sinuses but has little effect on formation. If occlusion does not promptly increase CSF pressure throughout the system, this is called a positive Queckenstedt–Stokey sign of pathology. The arachnoid villi collapse when CSF pressure exceeds cerebral vein pressure.

309. (C) Plasma has 300 times the concentration of proteins as CSF and therefore is a better buffered solution. CSF formation is not dependent upon a high filtration pressure. It is formed by an active process. CSF tends to maintain a fairly constant transmural pressure during the Valsalva maneuver, during headward acceleration and deceleration, and when one changes from a horizontal to a vertical position. For example, in the standing position, while gravity is tending to pull blood caudad, it is also tending to pull CSF caudad. In this way, CSF tends to maintain the patency of the cerebral vessels. During headward deceleration, the CSF tends to prevent rupture of the cerebral blood vessels. CSF has a lower concentration of K^+ and Ca^{++} than plasma (0.62 times the concentration of K^+ and 0.49 times the concentration of Ca^{++}).

310. (E) When the cerebral aqueduct is blocked, the CSF pressure may change from a normal value of 100 mm H_2O (7 mm Hg) to one of 160 mm H_2O (12 mm Hg). In the fetus or newborn child, this may affect brain development and cause hydrocephalus. In the adult, it may cause brain ischemia.

311. (E) Intraocular fluid contains 18 times the concentration of ascorbic acid found in plasma and 12 times the concentration found in CSF. It has a lower protein concentration than either plasma or CSF. The pressure may rise from a normal value of 20 mm Hg to one in excess of 80 mm Hg. Pressures as high as 40 mm Hg can cause blindness. Intraocular fluid, like CSF, is formed by active transport. It is an important source of nutrient for the lens.

312. (D) Intraocular fluid helps maintain the curvature of the cornea. The canal of Schlemm is the major site of reabsorption. A structure in the ciliary process that is similar in structure and function to the choroid plexus is the major site of secretion.

313. (A) Any agent or procedure that produces dilation of the pupil tends to obstruct the flow of intraocular fluid into the canal of Schlemm. A procedure that increases the hydrostatic pressure in the capillaries of the corneoscleral junction of the eye will decrease absorption of intraocular fluid into the blood. Carbonic anhydrase inhibitors decrease the movement of HCO_3^-, Na^+, Cl^-, and, therefore, also water into the chambers of the eye. In other words, by interfering with active transport, it tends to reduce the pressure in the chambers of the eye.

314. (D) Endolymph is found in the scala media (cochlear duct) of the cochlea, ductus utriculosaccularis, ductus endolymphaticus, subdural sac, and the membranous labyrinths of the semicircular canals, utricle, and saccule.

315. (B) Perilymph is found in the scala vestibuli and scala tympani of the cochlea and in the osseous labyrinths of the semicircular canals, utricle, and saccule. The canaliculus cochleae is a connecting link between the perilymph of the cochlea and the subarachnoid space and may also contain perilymph.

316. (E)

Cerebrospinal fluid	0.21 mg of protein/mL
Intraocular fluid	0.10 mg of protein/mL
Plasma	72 mg of protein/mL
Amniotic fluid	2.5 mg of protein/mL
Endolymph	0.15 mg of protein/mL
Perilymph	0.50 mg of protein/mL
Urine	negligible

317. (C) Endolymph has been estimated to have a K^+ concentration of 154 mEq/L. The K^+ concentration in urine on an average is 60 mEq/L. The other fluids listed and plasma all contain a K^+ concentration below 9 mEq/L.

318. (B) Amniotic fluid serves no role in gas exchange. This function is reserved for the placenta.

CHAPTER 13

Hemostasis

Questions

DIRECTIONS (Questions 319 through 328): Each of the numbered items or incomplete statements in this section is followed by answers or by completions of the statement. Select the ONE lettered answer or completion that is BEST in each case.

319. All of the following are important responses of the body that prevent blood loss after the rupture of a small blood vessel EXCEPT

(A) formation of a platelet plug
(B) hypotension, decreasing perfusion pressure
(C) vasoconstriction
(D) an increased perivascular pressure associated with the production of a hematoma
(E) formation of insoluble fibrin threads

320. Which of the following is released by blood platelets during hemorrhage and tends to produce vasoconstriction?

(A) serotonin
(B) histamine
(C) thrombosthenin
(D) accelerator globulin
(E) bradykinin

321. What function(s) do platelets serve?

(A) release of a vasoconstrictor
(B) they are essential for clot retraction
(C) formation of a white thrombus (ie, a plug)
(D) through the release of platelet factors, they initiate and accelerate the formation of a fibrin clot
(E) all of the above

322. Which of the following statements is MOST correct? A procoagulant not normally circulating in the plasma is

(A) prothrombin
(B) fibrinogen
(C) antihemophilic factor (factor VIII)
(D) Ac-globulin (factor V or proaccelerin)
(E) none are correct

323. Dicumarol is a drug that impairs the utilization of vitamin K by the liver. Dicumarol therapy, therefore, would decrease the plasma concentration of which of the following procoagulants?

(A) prothrombin
(B) fibrinogen
(C) antihemophilic factor (factor VIII)
(D) Ac-globulin (factor V)
(E) none of the above

324. Which of the following combinations of substances in plasma will cause the production of a clot?

(A) prothrombin, accelerator globulin, antihemophilic factor, platelet factor 3, fibrinogen
(B) thrombin and fibrinogen
(C) prothrombin, accelerator globulin, tissue extract, Ca^{++}
(D) accelerator globulin, antihemophilic factor, platelet factor 3, Ca^{++}, fibrinogen
(E) prothrombin, accelerator globulin, fibrinogen, Ca^{++}

325. Rheomacrodex (low-molecular-weight dextran) is administered to a patient and found to have no effect on his recalcification plasma clotting time or one-stage prothrombin time but does increase his bleeding time. Which of the following statements is the most likely explanation of this action? Rheomacrodex

(A) inhibits platelet aggregation
(B) inhibits rouleau formation
(C) inhibits fibrin polymerization
(D) inhibits the action of thrombin
(E) facilitates the activation of factor X

326. Which of the following statements is correct? Serum does NOT contain

(A) prothrombin
(B) plasma thromboplastin component (PTC, factor IX)
(C) Ca^{++}
(D) serum prothrombin conversion accelerator (SPCA, factor VII)
(E) any of the above

327. You have a patient in whom you wish to prolong the Lee–White clotting time for the next 10 hours. What would be the procedure of choice?

(A) an intravenous drip of a heparin solution
(B) an intravenous injection of sodium citrate
(C) the administration of dicumarol
(D) any of the above would be effective; one would make a choice on the basis of cost, availability, and facilities
(E) intravenous vitamin K

328. Which of the following statements is correct?

(A) all coagulopathies (deficiencies in one of the 13 clotting factors) cause an increased Lee–White whole-blood clotting time
(B) all coagulopathies cause an increased one-stage prothrombin time
(C) all thrombocytopathies (platelet dysfunctions) are associated with a prolonged clot retraction time
(D) a prolonged Lee–White whole-blood clotting time is usually associated with a prolonged bleeding time
(E) none are correct

Answers and Explanations

319. (B) Hypotension may be a direct effect of blood loss, but re-establishment of normal blood pressure by various mechanisms is a normal response.

320. (A) Histamine produces vasodilation in the normal systemic arteriole. Thrombosthenin is a contractile substance in the platelet that causes clot retraction.

321. (E) Possibly, they are also deposited in the capillary membrane, where they contribute to its integrity.

322. (E) All of the factors listed are procoagulants found in plasma.

323. (A) Avitaminosis K causes either a decreased plasma concentration or a decreased production of the following factors during coagulation:

- prothrombin
- serum prothrombin conversion accelerator (SPCA, factor VII, autoprothrombin I)
- plasma thromboplastin component (PTC, factor IX, Christmas factor, antihemophilic factor B, autoprothrombin II)
- Stuart–Prower factor (factor X, autoprothrombin III)

These are all procoagulants that are either produced by the liver or whose precursors are produced by the liver.

324. (B) Thrombin catalyzes the conversion of fibrinogen to fibrin, a monomer. Fibrin monomers undergo end-to-end and side-to-side polymerization to form extensive networks of fibrin threads.

325. (A) Both the recalcification plasma clotting time and the prothrombin time are performed on plasma (ie, in the absence of platelets). The fact that these two tests yield normal results is indicative of a normal clotting system. The fact that the bleeding time is prolonged is indicative of a failure in some other aspect of the hemostatic mechanism. There is no evidence to indicate that the tendency of red cells to form a rouleau affects the bleeding time. Factor X (Stewart factor) catalyzes the formation of thrombin.

326. (A) Serum is the solution released from a blood clot during clot retraction. Normally, it is devoid of prothrombin, fibrinogen, and factors V (Ac-globulin), VIII (antihemophilic factor), and XI (plasma thromboplastin antecedent [PTA]). Apparently, these substances are consumed in the clotting reaction. PTC, Ca^{++}, and SPCA all either catalyze or facilitate the formation of thrombin without being completely consumed. Factors X, XII, and XIII are also found in serum.

327. (A) An intravenous drip of a heparin solution would be the procedure of choice.

328. **(E)** Deficiencies in fibrin-stabilizing factor (factor XIII) and serum prothrombin conversion accelerator (SPCA, factor VII) do not cause a prolonged Lee–White clotting time. In factor VII deficiency, however, the prothrombin time is prolonged. Deficiencies in factors VIII, IX, XI, XII, and XIII do not produce a prolonged prothrombin time. This does not mean that all of these procoagulants are unimportant in the activation of prothrombin but that in a prothrombin time procedure an excess of thromboplastin is added to the sample. This excess will mask the action of the above factors. In hereditary, hemorrhagic thrombasthenia, clot retraction is either absent or prolonged. In von Willebrand's syndrome (an inherited thrombocytopathy), clot retraction is normal. In most coagulopathies the bleeding time is normal. A prolonged bleeding time with a normal whole-blood clotting time and prothrombin time is usually caused by a thrombocytopathy.

CHAPTER 14

Blood

Questions

DIRECTIONS (Questions 329 through 344): Each of the numbered items or incomplete statements in this section is followed by answers or by completions of the statement. Select the ONE lettered answer or completion that is BEST in each case.

329. Which of the following statements about albumin is correct?

(A) less than 10% is degraded each month
(B) it has a molecular weight greater than that for the gamma globulins
(C) it is responsible for most of the osmotic pressure of plasma
(D) in plasma, it approaches the concentration of the plasma globulins in liver disease and nephrosis
(E) at a pH of 7.4, it is a positively charged particle

330. The plasma proteins function to (1) destroy certain foreign materials (antibodies), (2) control body activity (hormones), and (3) prevent bleeding (procoagulants). They serve the other functions listed EXCEPT

(A) they prevent edema by causing the plasma osmotic pressure to be higher than the perivascular osmotic pressure
(B) they provide 75% of the buffering capacity of the blood
(C) they prevent excretion and destruction of hormones, vitamins, iron, lipids, drugs, etc. The plasma proteins do this by forming complexes with these substances. For example, a 6-year-old girl with a transferrin (a β_1 globulin) deficiency had a plasma iron-binding capacity of 15 μg/100 mL (330 μg/100 mL is normal) and a disappearance time for iron of 5 minutes (70 to 140 minutes is normal)
(D) they decrease the potency of agents (thyroxine, for example) by forming protein complexes that serve as reservoirs but are not themselves active
(E) they lyse clots (profibrinolysin)

331. Which of the following statements concerning antibodies is INCORRECT?

(A) they are all proteins
(B) they are produced by the ribosomes in the cells of the spleen and lymph nodes
(C) some cause lysis of antigen-containing cells
(D) some cause precipitation of an antigen
(E) they have a half-life of less than 24 hours

332. Which of the following statements concerning antibodies is INCORRECT?

(A) the fetus in the 8th month of pregnancy has about one-half the antibody-producing potential the adult has
(B) antibody production is more rapid and more marked after the second exposure to an antigen than after a first exposure 1 month earlier
(C) a high antibody titer is usually associated with a high lymphocyte count
(D) antibody production is depressed by adrenocorticotropic hormone (ACTH), radiation, and nitrogen mustards
(E) antigen–antibody reactions are associated with the release of histamine

333. Which of the following statements about lymphocytes is INCORRECT?

(A) they are produced by the thymus, red bone marrow, spleen, and lymph nodes
(B) their concentration in the blood falls abruptly after removal of the thymus gland in the adult, and the immune reaction is disturbed
(C) they are capable of changing into plasma cells which secrete antibodies
(D) they constitute 20 to 40% of the leukocytes
(E) they do not perform an important phagocytic function

334. Which of the following statements concerning the monocyte is INCORRECT?

(A) it is more common in the blood than the eosinophil and basophil
(B) it is produced in the adult by the bone marrow and lymph nodes
(C) unlike the neutrophil, it is rich in lipase
(D) unlike the neutrophil, it does not accumulate outside the circulation in an area of inflammation
(E) it is not classified as a granulocyte

335. Which of the following statements is correct? The neutrophil

(A) is the second most numerous leukocyte in the blood
(B) is not actively phagocytic in the bloodstream
(C) has a life span of about 120 days
(D) acts as a reservoir for vitamin B_{12}
(E) is formed in the adult mainly in the spleen and lymph nodes

336. Which of the following statements is INCORRECT? Neutrophilic granulocytes

(A) contain histamine
(B) are in the blood at a concentration of 200,000 to 500,000/mm^3
(C) are attracted to a site of tissue damage by chemical agents such as histamine
(D) contain lysosomes
(E) increase in concentration during exercise

337. Which of the following statements is INCORRECT? The concentration of neutrophils in the blood

(A) rises following a cardiac infarction (necrosis of an area in the heart)
(B) rises whenever the lymphocyte count rises
(C) rises whenever an infection is associated with the formation of pus
(D) falls in response to hypoxia
(E) falls during vitamin B_{12} deficiency

338. There is still a great deal that we do not know about the function of the leukocytes. Which of the following perspectives is LEAST likely to be correct?

(A) basophils are granulocytes that contain histamine and heparin in their metachromatic granules
(B) eosinophils phagocytose the antigen–antibody complex
(C) the eosinophil concentration in the blood increases during allergic reactions

(D) eosinophils are sites for the manufacture of antibodies
(E) neutrophils have multilobed nuclei

339. Which of the following statements is INCORRECT?

(A) antigens are not always proteins
(B) in some cases, antigens require the presence of a plasma factor (complement) before they will combine with an antibody
(C) the body can form antibodies against substances with a molecular weight less than 10,000
(D) the antigens responsible for the A and B blood groups are unconjugated proteins
(E) the antigen responsible for Rh positivity has been called the D factor by Fisher

340. Blood for transfusion is stored at 4°C for periods up to 21 days. It usually has had sodium citrate (prevents clotting), citric acid (reduces pH), and dextrose (additional source of nutrient) added to it. Which of the following changes would NOT occur in the blood after 7 days of storage?

(A) a decreased concentration of dextrose and an increased concentration of lactic acid
(B) a decreased concentration of plasma K^+
(C) an increased prothrombin time
(D) a decreased concentration of SPCA (factor VII), antihemophilic factor (AHF, factor VIII), and PTC (factor IX)
(E) a decreased concentration of platelets

341. Which one of the following statements is correct? An individual with blood group A1

(A) will not develop plasma agglutinins to type B blood in the absence of exposure to a B agglutinogen
(B) will never agglutinate the erythrocytes from a group A1 donor
(C) usually will agglutinate the erythrocytes from a group O donor
(D) usually will agglutinate the erythrocytes from a group AB donor
(E) usually will agglutinate the erythrocytes from a type A2 donor

342. Which of the following statements is correct? An individual with blood group B

(A) has the most common blood type
(B) cannot be the biologic father of a child with an O blood type
(C) cannot be the biologic father of a child with an A blood type
(D) cannot be the biologic father of a child with an AB blood type if the mother is type O
(E) all are correct

343. Which of the following represents the MOST potentially dangerous situation?

(A) an Rh-positive mother who is bearing her second Rh-negative child
(B) an Rh-negative mother who is bearing her second Rh-positive child
(C) an Rh-positive mother who is bearing her first Rh-negative child
(D) an Rh-negative mother who is bearing her first Rh-positive child
(E) the transfusion of 1 unit of O-negative blood into an A-positive recipient

344. What is the MOST appropriate treatment for a newborn child (blood group A) suffering from erythroblastosis fetalis?

(A) simultaneous bleeding and transfusion with A-positive blood
(B) simultaneous bleeding and transfusion with A-negative blood
(C) simultaneous bleeding and transfusion with either A-positive or A-negative blood
(D) transfusion with type O blood
(E) transfusion with type AB blood

Answers and Explanations

329. (D) Normally, the albumin–globulin (A-G) ratio is 2:1. In liver disease, albumin production decreases. In nephrosis, the loss of albumin increases (albuminuria). Since these conditions are not associated with a similar decrease in globulin, but may cause an increased plasma globulin concentration, the A-G ratio may change from 2:1 to 1:1. Approximately 6 to 10% of the albumin in plasma is degraded each day. The molecular weight for albumin is approximately 69,000 and that for gamma globulin is 156,000. The osmotic pressure of plasma is 5776 mm Hg. Albumin normally is responsible for 16.4 mm Hg pressure and the other colloids about 9 mm Hg. Albumin is negatively charged at a pH of 7.4.

330. (B) The plasma proteins provide one-sixth of the buffering capacity of the blood.

331. (E) It is the plasma cells in the spleen and lymph nodes that produce the antibodies. The average biological half-life of antibodies is about 13 days in the adult.

332. (A) At birth, the child has little or no antibody-producing potential. A first exposure to an antigen results in a maximum antibody titer in 8 to 12 days. A second exposure results in a maximum titer that is higher than occurred initially and is seen in 5 to 9 days. Small lymphocytes produce and carry antibodies. ACTH facilitates the release of cortisol, which, in turn, depresses antibody production and decreases the number of circulating lymphocytes. The release of histamine accounts for some of the symptoms found in allergic reactions.

333. (B) The thymus is smaller and of less importance in the adult than in the child. In the child, the thymus is an important source of lymphocytes. It has been suggested that in the newborn mouse, the thymus is essential for the development of immune responses, but these observations have not been confirmed on pathogen-free animals. In the fetus, lymphocyte precursors are changed by the thymus to T lymphocytes and these are changed by the liver and spleen to B lymphocytes. Both T and B lymphocytes migrate to the lymph nodes and bone marrow, where they function in the child and adult in cellular and humoral immunity.

334. (D) Both monocytes and neutrophils rapidly accumulate at a site of infection, with the neutrophils predominating initially. The neutrophils disintegrate within about 3 days and then the monocytes and lymphocytes predominate. Monocytes constitute 4 to 8% of the leukocytes, eosinophils 1 to 3%, and basophils 0 to 1%.

335. (D) The neutrophil represents 50 to 70% of the leukocytes in the blood. It is phagocytic both in and out of the blood. In the blood it phagocytoses bacteria, small insoluble particles, and fibrin. Its life span is less than 2 weeks. It is formed mainly in the red bone marrow.

336. (B) There are normally 3000 to 6000 neutrophils per mm^3 of blood. It has been suggested that the release of histamine by neutrophils serves to increase the circulation to an area (vasodilation) and serves as a chemical signal to bring to an area of infection more neutrophils (chemotaxis).

337. (B) The movement of lymphocytes and neutrophils into the blood can be independently controlled. Adrenocorticotropic hormone will, for example, increase the number of circulating neutrophils while it decreases the number of circulating lymphocytes and eosinophils. The products of tissue damage facilitate the movement of neutrophils from reservoirs into the blood. Pus consists primarily of dead neutrophils. Hypoxia also causes an increase in erythrocyte production as a result of an increase in erythropoietic activity in the bone marrow and possibly also as a result of a decrease in leukopoiesis.

338. (D)

339. (D) The antigens responsible for the A and B blood groups are mucopolysaccharides. Antigens are usually proteins, polysaccharides, or combinations. Characteristically, the body forms antibodies against substances with a molecular weight in excess of 10,000, but molecules with a molecular weight of less than this can combine with an antigen and, in this way, stimulate the production of antibodies that will react against the low-molecular-weight substance (called a hapten) when it is not in combination with another molecule.

340. (B) The plasma K^+ will change from 3.5 mEq/L to 12 mEq/L. Other signs of deterioration of cellular function include an increased plasma hemoglobin (from 0 to 10 to 25 mg/100 mL), plasma inorganic phosphate (1.8 to 4.5 mg/100 mL), and plasma ammonia (50 to 260 μg/100 mL). The dextrose will change from a plasma concentration of 350 mg/100 mL to 300 mg/100 mL, and the lactic acid will change from 20 mg/100 mL to 70 mg/100 mL. The prothrombin time increases by 50%. The plasma concentrations of the following decrease: SPCA (–11%), AHF (–62%), PTC (–18%). The platelet count decreases 52%.

341. (D) The A1 recipient contains anti-B agglutinins. The anti-B agglutinin is always found in the plasma of individuals with blood group A1, A2, or O. There is no evidence that it develops in response to the exposure of the individual to a B agglutinogen. If the A1 recipient is Rh-negative and has previously received Rh-positive whole blood, he may agglutinate Rh-positive A1 cells. Physicians, by using the ABO classification, can reduce transfusion deaths to 2 per 1000 transfusions. In addition, by using the Rh classification, they can still further reduce transfusion deaths. The group O donor's erythrocytes lack the A and B agglutinogens. The A1 recipient generally does not contain anti-A2 agglutinins.

342. (D) A child with an AB phenotype must have an AB genotype. If its father is type B, its mother must have an A or AB phenotype. The most common blood type among Caucasians (45%), Negroes (48%), and Orientals (36%) is group O. Group A ranks second (41%, 27%, 28%), group B third (10%, 21%, 23%), and group AB fourth (4%, 4%, and 13%). An individual with a group B phenotype can have a BB or BO genotype. A child with an A phenotype can have either an AA or AO genotype.

343. (B) The fetus is incapable of producing dangerous quantities of antibodies against its mother's cells. On the other hand, an Rh-negative mother can produce dangerous quantities of antibodies against the cells of an Rh-positive fetus. An Rh-negative mother who has not been previously exposed to Rh-positive cells will have no agglutination problem during a first Rh-positive pregnancy. Only 1 in 42 have problems in their second pregnancy and only 1 in 12 in their fifth pregnancy. In other words, the incidence of erythroblastosis fetalis will depend on the number of exposures of the mother to Rh-positive cells, and this will depend, in part, on the number of pregnancies and the adequacy of the placental barrier. The transfusion of 1 unit of O-negative blood into an A-positive recipient is a safe procedure.

344. **(B)** Because the Rh-positive infant shows little or no reaction against Rh-negative cells, this is the procedure of choice. The newborn with hemolytic disease has two problems: (1) too many antibodies to his Rh-positive cells and (2) Rh-positive cells. Transfusion with Rh-positive blood tends to dilute the antibodies but does nothing about the number of Rh-positive cells. Type O-negative would be an acceptable transfusion. Transfusion of the infant with type AB blood would be counterproductive.

CHAPTER 15

The Erythrocyte

Questions

DIRECTIONS (Questions 345 through 372): Each of the numbered items or incomplete statements in this section is followed by answers or by completions of the statement. Select the ONE lettered answer or completion that is BEST in each case.

345. During the 3rd, 4th, and 5th months of gestation, fetal hemoglobin is produced primarily in which one of the following structures in the fetus?

(A) blood islands
(B) red bone marrow
(C) spleen
(D) liver
(E) lymph nodes

346. In a healthy 40-year-old male, red blood cells are formed in the

(A) femur
(B) tibia
(C) vertebrae
(D) A and C
(E) A, B, and C

347. The principal site of production of the erythropoietic factor (ie, erythrogenin) is thought to be the

(A) red bone marrow
(B) spleen
(C) lymph nodes
(D) kidney
(E) small intestine

348. Which of the following statements concerning the red blood cell (RBC) and hemoglobin (Hb) is correct? Under healthy conditions the blood contains approximately

(A) 5×10^6 RBC per mL and 15 g of Hb per mL
(B) 5×10^6 RBC per mL and 0.15 g of Hb per mL
(C) 5×10^6 RBC per mL and 0.15 mg of Hb per mL
(D) 5×10^6 RBC per mm^3 and 0.15 mg of Hb per mL
(E) 5×10^6 RBC per mm^3 and 0.15 g of Hb per mL

349. The following data are collected during a study on a patient: Total volume of air expired in 5 minutes: 100 liters at standard temperature, pressure, and dryness (STPD)

Pressure	760 mm Hg
O_2 in expired gas	17% (STPD)
CO_2 in expired gas	2.5% (STPD)
O_2 in inspired gas	20% (STPD)

What would the partial pressure of O_2 be in the expired gas at STPD?

(A) 99 mm Hg
(B) 109 mm Hg
(C) 119 mm Hg
(D) 129 mm Hg
(E) 139 mm Hg

350. What would the oxygen consumption of the patient in Question 349 be?

(A) 0.2 L/min
(B) 0.4 L/min
(C) 0.6 L/min
(D) 0.8 L/min
(E) 1.0 L/min

351. What would the respiratory exchange ratio (R) for the patient in Question 349 be?

(A) 1.02
(B) 0.97
(C) 0.92
(D) 0.87
(E) 0.82

352. A subject is breathing 50% O_2 and 50% N_2 at a barometric pressure of 760 mm Hg and room temperature. Which of the following statements is INCORRECT?

(A) $P_{N_2} = P_{O_2}$ in the air
(B) $P_{N_2} < 370$ mm Hg in the aortic blood
(C) $P_{O_2} > P_{N_2}$ in the aortic blood
(D) the O_2 content of the aortic blood exceeds the N_2 content of the aortic blood
(E) the P_{O_2} in the aortic blood will be directly related to the atmospheric pressure

353. When the P_{O_2} of the blood is 100 mm Hg, how much O_2 is dissolved in the blood as O_2?

(A) less than 5% of the total O_2 that enters the blood
(B) less than 20%
(C) less than 40%
(D) less than 60%
(E) more than 60%

354. When the P_{O_2} of the blood changes from 100 to 600 mm Hg, there is

(A) a sixfold increase in the O_2 content of the blood
(B) less than a 1% increase in the O_2 content of the blood
(C) a twofold increase in the O_2 content of the blood
(D) a sixfold increase in the quantity of dissolved O_2
(E) a sixfold increase in the quantity of dissolved O_2 and a twofold increase in the O_2 content of the blood

355. If one increases the alveolar P_{O_2} from 75 mm Hg to 100 mm Hg, the systemic artery blood will carry

(A) one third more O_2
(B) one third less O_2
(C) 5% more O_2
(D) 5% less O_2
(E) 20% more O_2

356. If one decreases the plasma P_{O_2} from 50 mm Hg to 30 mm Hg, the blood from a normal subject will carry

(A) one third more O_2
(B) one third less O_2
(C) 5% more O_2
(D) 5% less O_2
(E) 20% more O_2

357. Curve H in the following diagram (Fig. 15–1) represents a curve for normal arterial blood in a resting subject breathing room air. Which of the following statements is correct?

(A) curve F is for arterial blood from a subject who is hyperventilating
(B) curve G is for arterial blood from a subject with anemia

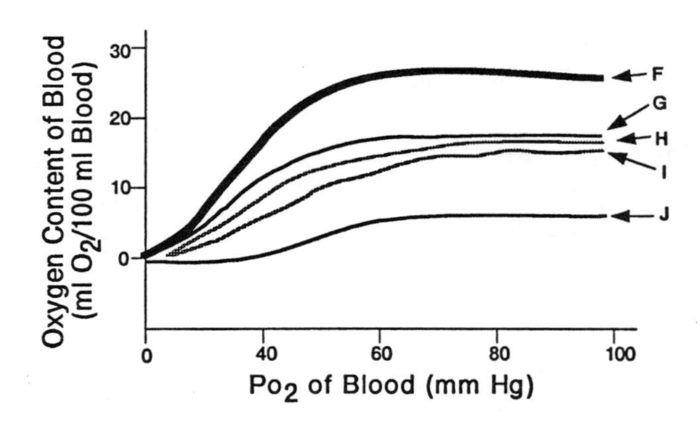

Figure 15–1. Oxygen dissociation curve for blood.

(C) curve I is for arterial blood from a subject with anemia
(D) curve I is for blood leaving an exercising muscle
(E) curve J is for blood leaving an exercising muscle

358. The concentration of O_2 in mL of O_2/100 mL of blood held at a PO_2 of 40 mm Hg is increased by

(A) an increased plasma pH from 7.4 to 7.6
(B) a decreased plasma pH from 7.6 to 7.4
(C) an increased plasma PCO_2 from 40 to 80 mm Hg
(D) an increased body temperature from 38 to 40°C
(E) B and D

359. Arterial PO_2 will be highest in a normal healthy adult under which of the following conditions?

	Alveolar Ventilation (L/min)	Blood Hemoglobin Concentration (g/dL)	PO_2 of Inspired Air (mm Hg)
(A)	8	8	90
(B)	7	8	300
(C)	7	18	150
(D)	4	8	90
(E)	4	21	150

360. The P50 for blood is the PO_2 at which the hemoglobin is 50% saturated with O_2. Under resting conditions, it is about 26 mm Hg. Which of the following statements is correct?

(A) 2,3-diphosphoglycerate (DPG) increases the P50
(B) DPG shifts the oxygen hemoglobin dissociation curve to the left
(C) fetal hemoglobin has a greater affinity for binding O_2 than does adult hemoglobin A because fetal hemoglobin contains less DPG
(D) acidosis, hypercapnea, and hypoxia decrease the concentration of DPG
(E) A, C and D

361. Cyanosis is caused by

(A) an increased concentration of unoxygenated hemoglobin
(B) a decreased concentration of hemoglobin
(C) a decreased concentration of oxyhemoglobin
(D) carbon monoxide poisoning
(E) hypoxia

362. Nitrites have which of the following effects?

(A) they reduce the ferric iron in hemoglobin to ferrous iron
(B) they increase the O_2 holding capacity of hemoglobin
(C) A and B
(D) they may produce hypoxia
(E) A and D

363. A healthy subject breathing air contaminated with CO (0.1% of the total volume) at a pressure of 760 mm Hg will have an arterial plasma

(A) PCO_2 that exceeds the venous plasma PCO_2
(B) PCO_2 that approximately equals the venous plasma PCO_2
(C) PCO that exceeds the arterial plasma PO_2
(D) PO_2 that exceeds the arterial plasma PN_2
(E) PN_2 that exceeds the arterial plasma PO_2

364. If the conditions in Question 363 are maintained, the subject dies. This is caused by

(A) the irritant action of CO on the respiratory membrane
(B) the direct action of CO upon the enzymes of the cell
(C) CO interfering with the action of carbonic anhydrase
(D) CO blocking the action of erythropoietin
(E) CO combining with hemoglobin

365. In which of the following conditions is the arterial Po_2 reduced?

(A) anemia
(B) cyanide poisoning
(C) pulmonary hypoventilation
(D) CO poisoning
(E) moderate exercise

366. Which of the following changes in the blood of the right atrium is most likely to be seen in anemia?

(A) an increase in O_2-carrying capacity
(B) a decrease in hemoglobin saturation
(C) an increase in O_2 content
(D) an increase in Po_2
(E) none of the above

367. Which of the following statements is INCORRECT?

(A) an individual who is homozygous for hemoglobin-S has hemoglobin that is insoluble at a low Po_2
(B) an individual who is heterozygous (ie, has one gene for Hb-S and one for Hb-A) has only normal hemoglobin
(C) an individual who has the sickle cell trait usually has some Hb-A in his red cells
(D) ten percent of the U.S. black population contains the sickle cell trait
(E) in sickle cell disease, the P50 is increased

368. When the pH of the blood changes from 7.5 to 6.5, 31.74 mEq of H^+ combine with blood proteins. It is estimated that 87% of this buffering action is due to

(A) the plasma proteins because of the buffering capacity of albumin
(B) the plasma proteins because of the buffering capacity of the globulins
(C) hemoglobin, because it is in higher concentration in the blood than the plasma proteins
(D) hemoglobin, because it has a greater buffering capacity per gram
(E) hemoglobin, because it is in higher concentration and has a greater buffering capacity

369. Approximately 90% of the CO_2 that enters the blood is carried as

(A) dissolved CO_2
(B) carbonic anhydrase
(C) carbamino CO_2
(D) carbonic acid
(E) bicarbonate

370. In arterial blood, the concentration of dissolved O_2 is 0.3 mL/100 mL of blood and the concentration of dissolved CO_2 is 2.4 mL/100 mL. Why is the concentration of dissolved CO_2 so much higher?

(A) in the alveolus, the concentration of CO_2 is higher than the concentration of O_2
(B) in the lungs, the diffusing capacity (mL of gas moved from the alveolus to the blood per minute per mm Hg difference in partial pressure) for CO_2 is greater than that for O_2
(C) most of the O_2 that enters the blood is combined with hemoglobin
(D) in plasma, the solubility (mL of gas dissolved/mL of solution) of CO_2 is greater than that for O_2
(E) in blood, CO_2 is in equilibrium with HCO_3^-

371. If you compare the Po_2–blood O_2 curve with the Pco_2–blood CO_2 curve (Po_2 and Pco_2 are on the abscissa), you find that

(A) both are relatively flat at partial pressures above 60 mm Hg
(B) at a given partial pressure, the gas content for O_2 and CO_2 is approximately the same

(C) at a partial pressure of between 10 mm Hg and 100 mm Hg, both curves have a straight-line relationship between pressure and content

(D) increases in Po_2 shift the CO_2 curve to the right and increases in Pco_2 shift the O_2 curve to the right

(E) increases in acidity shift the two curves to the left

372. When CO_2 is removed from the blood in the lungs, which one of the following events also occurs? There is a net

(A) efflux of HCO_3^- from the erythrocyte (RBC)

(B) efflux of Cl^- from the erythrocyte

(C) influx of CO_2

(D) efflux of O_2

(E) efflux of O_2 and influx of CO_2

Answers and Explanations

345. (D)

346. (C) The femur and tibia stop producing erythrocytes at about age 20).

347. (D) It is proposed that hypoxia acts to increase the release of renal erythropoietic factor (REF) from the kidney and a globulin from the liver. The REF acts in the blood on the globulin to change it to erythropoietin. The erythropoietin is apparently essential for the maintenance of normal red cell production. It is destroyed in the liver and has a half-life of about 5 hours.

348. (E) There are between 4.3 and 6.0×10^9 RBC/mL and 0.15 g of Hb/mL. Because 1 mL of blood weighs between 1.050 and 1.060 g, it cannot contain 15 g of Hb per mL.

349. (D) Partial pressure of O_2:

$$P_{O_2} = (\%\ O_2/100) \times (\text{pressure}) = (17/100) \times (760\ \text{mm Hg}) = 129.2\ \text{mm Hg}$$

350. (C)

$$\text{Oxygen consumption} = (\%\ O_2\ \text{inspired} - \%\ O_2\ \text{expired})/100 \times (\text{volume expired/min}) = (20 - 17)/100 \times (100\ \text{L}/5\ \text{min}) = 0.6\ \text{L/min}$$

351. (E)

$$R = (CO_2\ \text{production})/(O_2\ \text{consumption})$$

$$CO_2\ \text{production} = (\%\ CO_2\ \text{expired} - \%\ CO_2\ \text{inspired})/100 \times (\text{volume expired/min}) = (2.5 - 0)/100 \times (100\ \text{L}/5\ \text{min}) = 0.5\ \text{L/min}$$

$$R = (0.5\ \text{L/min})\ (0.6\ \text{L/min}) = 0.83$$

Note: In doing this problem, it is assumed the inspired CO_2 concentration is approximately 0%. Since it is usually less than 0.4% this is a good approximation for this problem.

352. (C) The P_{N_2} will be greater than the P_{O_2}. This is because the inspired air will be diluted with an O_2 poor residual volume. The inspired air will be diluted with H_2O vapor (6.2% of the alveolar air; 47 mm Hg) and the residual volume of air in the respiratory system. The dilution of the air with H_2O will reduce the P_{N_2} from 380 mm Hg to 356.5 mm Hg [380 –(47/2 mm Hg)]. The rest of the dilution will produce only minor changes in the P_{N_2}. Hemoglobin markedly increases the O_2 content of the arterial blood. In a healthy subject, the aortic P_{O_2} = (% concentration in alveoli/100) (atmospheric pressure).

353. (A) At this P_{O_2}, each 100 mL of blood will contain 19.5 mL of O_2 combined with hemoglobin and 0.3 mL dissolved as free O_2.

354. (D) At a P_{O_2} of 100 mm Hg, the major carrier of O_2 in the blood is about 97.5% saturated. Therefore, much of the increase in blood O_2 content gained by increasing the P_{O_2} of the blood to 600 mm Hg will be gained by increasing the quantity of dissolved O_2.

PO_2	100 MM HG	600 MM HG
Dissolved O_2 in blood	0.3 mL/100 mL*	1.8 mL/100 mL
O_2 combined with Hb	19.5 mL/100 mL	20.8 mL/100 mL

* mL of O_2 per 100 mL of blood

355. (C) The relationship between the quantity of O_2 carried by hemoglobin (ordinate) and the PO_2 (abscissa) produces an "S"-shaped curve. Since the hemoglobin is about 94% saturated with O_2 at a PO_2 of 75 mm Hg, the amount of additional O_2 that will be carried by the blood in healthy subjects above this alveolar PO_2 is relatively small.

356. (B) The steepest part of the oxyhemoglobin dissociation curve is between a PO_2 of 10 and 50 mm Hg.

PLASMA PO_2	10 MM HG	30 MM HG	50 MM HG
Whole blood O_2	2.73 mL*	11.49 mL	16.85
Dissolved O_2	0.03 mL	0.09 mL	0.15 mL
Hemoglobin O_2	2.70 mL	11.40 mL	16.70 mL
Hemoglobin saturation	13.5%	57%	83.50%

* mL of O_2 at STPD per 100 mL of blood

357. (D) Metabolically active muscle produces heat, CO_2, and H^+. Each of these metabolites shifts the curve to the right at PO_2s below 80 mm Hg. Hyperventilation can markedly increase arterial PO_2, but not O_2 content at a particular PO_2. Curve J in Figure 15–1 is from an individual with anemia. In anemia the problem is a low O_2 content at each PO_2.

358. (A) A decrease in acidity shifts the PO_2–hemoglobin saturation curve to the left, increasing hemoglobin's affinity for oxygen. Increases in PCO_2 and increases in temperature will decrease the amount of O_2 carried by the blood.

359. (B) In the absence of pulmonary shunts, blood hemoglobin concentration plays no role in determining the PO_2 of arterial blood. Since only about 2% of the blood bypasses the pulmonary capillaries in a normal individual, the effect of hemoglobin on arterial PO_2 is negligible. On the other hand, a low hemoglobin may cause a reduction in venous PO_2. (B) is a better answer than (A) because the high PO_2 in (B) more than compensates the somewhat lower ventilation.

360. (E) Hemoglobin, at a high DPG level, will be 50% saturated with O_2 at a PO_2 greater than 26 mm Hg. DPG, by shifting the curve to the right, increases the amount of O_2 liberated from the blood at the PO_2 found in the perivascular spaces of the systemic capillaries. They increase the concentration of DPG.

361. (A) Cyanosis is first noted when the concentration of unoxygenated hemoglobin exceeds 5 g/100 mL of blood. Iron deficiency anemia, as well as other forms of anemia, can produce marked decreases in the concentration of hemoglobin and oxyhemoglobin without producing blueness. Carbon monoxide poisoning, as well as anemia, tends to prevent cyanosis by decreasing the concentration of unoxygenated hemoglobin. When hypoxia is caused by anemia, KCN, or CO, there is generally no cyanosis.

362. (D) Nitrites oxidize the ferrous iron to ferric iron. Nitrites decrease the O_2 holding capacity of hemoglobin. This represents the major problem. The methemoglobin formed by the oxidation of the ferrous iron is an ineffective O_2-carrying system.

363. (E) Because CO_2 is added to the blood in the systemic capillaries, the systemic vein blood will have a higher PCO_2 than the systemic artery blood. Under steady state conditions, the partial pressure of gases in the alveolus will equal the partial pressure of gases in the systemic artery blood. In other words, the arterial PO_2 will exceed the PCO and PCO_2 and be less than the PN_2.

364. (E) Direct action upon the enzymes of the cell is the mode of action of cyanide, but not CO. CO, under the conditions specified, combines with 50% of the O_2 receptor sites of hemoglobin and, in this way, halves the O_2 carrying capacity of the blood.

365. **(C)** In the healthy subject, the Po_2 of the blood leaving the lungs is equal to that in the alveoli. A. The arterial Po_2 is the concentration of O_2 in plasma expressed in mm Hg. It is not controlled by the concentration of hemoglobin or the number of red blood cells. Cyanide prevents the cell from utilizing O_2. It does not affect the arterial PO. Carbon monoxide ties up O_2 binding sites on hemoglobin. It does not affect the arterial Po_2. In the healthy subject, arterial Po_2 does not change appreciably during moderate exercise. Venous Po_2, on the other hand, is lowered.

366. **(B)** A decreased hemoglobin saturation with O_2 is not seen in the blood of the systemic arteries of the patient with uncomplicated anemia, but because the cells of the body continue to withdraw large quantities of O_2 from the blood, it will be seen in the systemic veins due to the decreased O_2 content in the systemic artery blood. In the anemic patient, there is a decreased right arterial blood O_2 content and partial pressure.

367. **(B)** Half of this individual's hemoglobin will be Hb-S and half Hb-A. At a low Po_2, Hb-S crystallizes within the erythrocyte to cause it to change from a flexible, biconcave to a fragile, rigid, crescent-shaped structure that tends to cause local ischemia. This individual is heterozygous for Hb-S. Because his erythrocytes usually contain combinations of Hb-S and Hb-A, they become less rigid during hypoxia and less likely to occlude the microcirculation than the erythrocytes from an individual with sickle cell anemia.

368. **(E)** There are normally about 3.9 g of plasma protein and 15 g of hemoglobin per 100 mL of blood.

369. **(E)** In the presence of the catalyst, carbonic anhydrase, CO_2 and H_2O are rapidly converted to H_2CO_3. The H_2CO_3 ionizes to H^+ and HCO_3^-. These reactions occur in areas where CO_2 is being produced. The reverse occurs where CO_2 is being lost (ie, in the lungs).

370. **(D)** This question is concerned with mechanisms (causes and effects). Statements (B) through (E) are true, but only one answer is the cause of the higher concentration of CO_2. Only one is causally related to the higher concentration of CO_2. CO_2 is about 20 times more soluble in plasma than O_2. CO_2 constitutes about 5% and O_2 about 13% of the alveolar air. In normal subjects, the blood is in the pulmonary capillaries long enough to come into equilibrium with the alveolar air. On the other hand, when there is a diffusion block (pulmonary edema, for example) the lower rate of diffusion for O_2 may result in a hypoxia not associated with hypercapnea. An equilibrium between alveolar O_2, dissolved O_2, and oxyhemoglobin is reached in the pulmonary capillaries, and therefore, the affinity of hemoglobin for O_2 does not change. The equilibrium between CO_2 and HCO_3^- is reached before the blood leaves the pulmonary capillaries.

371. **(D)** Carbaminohemoglobin ($HbCO_2^-$) and acid hemoglobin (HHb) have less affinity for O_2 than hemoglobin (Hb^-). Oxyhemoglobin (HbO_2^-) and acid hemoglobin (HHb) have less affinity for O_2 than hemoglobin (Hb^-). The CO_2 curve is not flat at a Pco_2 above 60 mm Hg. At a Pco_2 of 40 mm Hg, 49 mL of CO_2 are being carried by each 100 mL of blood as HCO_3^-, carbamino-CO_2, etc.; at a Po_2 of 40 mm Hg, 15 mL of O_2/100 mL of blood. Neither curve has a straight-line relationship at the pressures stated in choice (C). Acidity shifts the two curves to the right.

372. **(B)** Associated with the influx of negatively charged bicarbonate ions is the efflux of negatively charged chloride ions. This is called the chloride shift and serves to maintain a relatively constant transmembrane potential. Bicarbonate ions diffuse into the RBC as its carbonic anhydrase catalyzes the formation of CO_2 and H_2O from H_2CO_3. In the lungs, the RBC is producing CO_2 and the CO_2 is diffusing out into the plasma. Since the Po_2 in the alveolus is greater than the Po_2 in the venous blood entering the pulmonary capillaries, the Po_2 of the plasma will be initially greater than the Po_2 of the RBC cytoplasm. Therefore, there is an influx of O_2.

PART V

Respiration

CHAPTER 16

Ventilation

Questions

DIRECTIONS (Questions 373 through 400): Each of the numbered items or incomplete statements in this section is followed by answers or by completions of the statement. Select the ONE lettered answer or completion that is BEST in each case.

373. Normal inspiration results from

(A) a decreased intrapleural pressure
(B) an increased alveolar pressure
(C) A and B
(D) depression of the thorax
(E) relaxation of the diaphragm

374. During the initial phase of inspiration in a healthy subject at rest,

(A) intrapulmonary pressure rises
(B) intra-abdominal pressure rises
(C) intrapulmonary and intra-abdominal pressures rise
(D) there is less muscular effort than during the initial phase of expiration
(E) the larynx is elevated

375. Which of the following statements is correct?

(A) the respiratory bronchioles contain cartilage that maintain their patency
(B) dry air at 0°C that is inspired through the nasal cavity does not reach body temperature and 100% humidity until after it reaches the lobar bronchus
(C) the nasal and pharyngeal epithelium and trachea are devoid of functional cilia
(D) during quiet breathing, little or no mucus (less than 1 mL per hour) is produced in the respiratory tract
(E) during a quiet inspiration, there is a submaximal contraction of the diaphragm and external intercostal muscles

376. A patient is taken to surgery and given a general anesthetic and muscle relaxant. An anesthetist is responsible for maintaining an appropriate depth of anesthesia. Under these conditions, what is most likely responsible for each of the following events?

(A) a movement inward during inspiration of the intercostal tissue due to a paralysis of the external intercostal muscles
(B) a movement inward during inspiration of the intercostal tissue due to a contraction of the internal intercostal muscles
(C) a movement inward during inspiration of the intercostal tissue due to a relaxation of the external and a contraction of the internal intercostal tissues
(D) an abdominal breathing due to the paralysis of the diaphragm
(E) an abdominal breathing due to the contraction of the abdominal muscles

377. Under resting conditions, an individual inspires air about 6 L/min. When his ventilation exceeds 100 L/min, inspiration will be produced not only by the contraction of the diaphragm and external intercostal muscles, but also by the contraction of

(A) the scalene, sternomastoid, and trapezius muscles
(B) the scalene, sternomastoid, and transversus abdominis muscles
(C) the sternomastoid, transversus abdominis, and trapezius muscles
(D) the transversus abdominis, trapezius, and scalene muscles
(E) the transversus abdominis, trapezius, scalene, and sternomastoid muscles

378. The residual volume for the lungs

(A) is the volume of air that remains in the lungs after expiring the resting tidal volume of air
(B) is the volume of air remaining in the lungs after inspiring the resting tidal volume of air
(C) is generally greater at age 75 than at age 45
(D) is less than 0.5 L in the adult
(E) Increases in atelectasis

379. Which of the following formulae is correct?

(A) vital capacity = inspiratory reserve volume + expiratory reserve volume
(B) dead air space = resting tidal volume + residual volume
(C) alveolar minute ventilation = (respiratory rate) × (tidal volume − dead air space)
(D) vital capacity = inspiratory reserve volume + resting tidal volume + expiratory reserve volume + residual volume
(E) inspiratory reserve volume = vital capacity − resting tidal volume

380. An athlete at rest has a cardiac output of 4 L/min, an oxygen consumption of 0.25 L/min, and a pulmonary ventilation of 5 L/min. During running, his cardiac output increases to 20 L/min and his oxygen consumption to 3 L/min. Approximately what would his pulmonary ventilation be?

(A) 5 L/min
(B) 20 L/min
(C) 60 L/min
(D) 144 L/min
(E) 200 L/min

381. A subject is asked to breathe air through a tube. During the experiment the following data are collected:

Tidal volume	500 mL
Alveolar P_{CO_2}	40 mm Hg
P_{CO_2} of expired gas	30 mm Hg

What is the respiratory dead air space of this individual?

(A) less than 100 mL
(B) between 100 and 150 mL
(C) between 150 and 200 mL
(D) between 200 and 250 mL
(E) more than 250 mL

382. The anatomic dead space

(A) is the same at the functional residual capacity at total lung capacity
(B) represents approximately 5% of the functional residual capacity
(C) A and B
(D) represents approximately 5% of the total lung capacity
(E) and the physiologic dead space are determined by different methods but yield values in both normal and pathologic states that agree to within 10%

383. What might account for a decreased arterial P_{O_2} in a subject with a constant CO_2 production and constant respiratory minute volume?

(A) a decreased functional residual capacity
(B) a decreased respiratory rate and tidal volume
(C) an increased respiratory rate and a decreased tidal volume

(D) a decreased respiratory rate and an increased tidal volume
(E) an increased respiratory rate and tidal volume

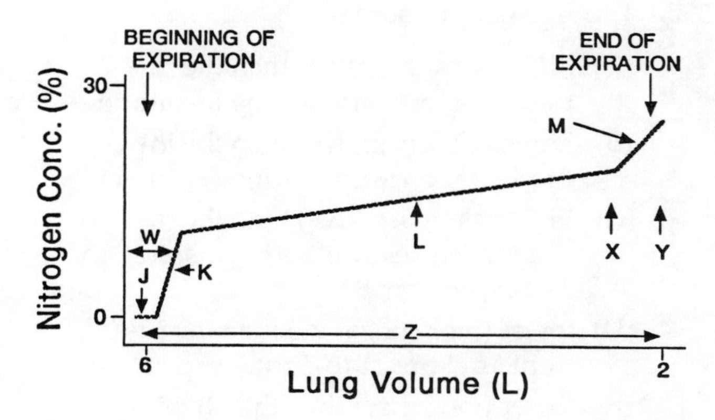

Figure 16–1. The N_2 content of a maximal exhalation by a subject following a deep inspiration in which 100% O_2 was inspired from midinspiration.

384. Which of the following statements best characterizes the results seen in Figure 16–1?

(A) the gas at J is primarily alveolar gas
(B) the gas at K is only dead space gas
(C) the gas at L is primarily dead space gas
(D) the volume of air moved during phase M is the dead space
(E) the high N_2 concentration in phase M is because the subject is expiring gas that comes primarily from the upper portions of the lungs

385. What conclusion can you draw from the loops in Figure 16–2? Subject X has the

(A) higher pulmonary compliance
(B) higher tidal volume
(C) higher pulmonary compliance and tidal volume
(D) lower pulmonary compliance
(E) lower pulmonary compliance and tidal volume

386. A unilateral pneumothorax causes which one of the following changes?

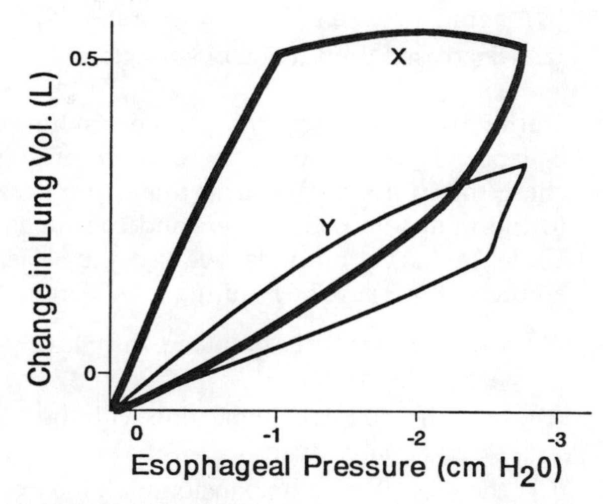

Figure 16–2. Pressure-volume loops obtained from subjects X and Y during quiet breathing at a rate of 14/min.

(A) an increase in pulmonary residual volume
(B) a collapse of the chest wall inward
(C) a decrease in the intrapleural pressure
(D) an increase in resting tidal volume
(E) a shift of the mediastinum toward the normal side

387. In young, healthy men the compliance for both the pulmonary system and the thoracic cage are each 0.2 L/cm of H_2O. What is the compliance of the respiratory system? (Assume that the respiratory system consists of only the thoracic cage and the pulmonary system.)

(A) 0.01
(B) 0.1
(C) 1.0
(D) 10
(E) 100

388. In which of the following conditions is there MOST likely to be an increased respiratory compliance and a decreased specific compliance?

(A) obesity and pregnancy
(B) diffuse alveolar fibrosis
(C) alveolar edema

(D) aging (between 45 and 90 years)
(E) decreased production of surfactant

389. During thoracic surgery, a patient's lungs collapse. Prior to the final closure of the chest, the surgeon applies air under pressure to the mouth in order to expand the lungs. Under what conditions would the least amount of pressure be required?

(A) the first 50 mL of expansion with the external nares open
(B) the third 50 mL of expansion with the external nares open
(C) the first 50 mL of expansion with the external nares closed
(D) the third 50 mL of expansion with the external nares closed
(E) the first 50 mL of expansion in the absence of surfactant production

390. The aspiration of amniotic fluid by transabdominal amniocentesis is proving a useful diagnostic tool. When the ratio of lecithin to sphingomyelin in the amniotic fluid is

(A) greater than 2, the fetus is immature and, if delivered, will probably suffer severe emphysema (dilation of alveoli)
(B) greater than 2, the fetus is immature and, if delivered, will probably suffer severe atelectasis (collapse of alveoli)
(C) less than 1, the fetus is immature and, if delivered, will probably suffer severe emphysema
(D) less than 1, the fetus is immature and, if delivered, will probably suffer severe atelectasis
(E) greater than 2, the fetus should be delivered by cesarean section

391. Which of the following statements is correct? Surfactant

(A) is distributed homogeneously throughout the liquid that covers the alveolar epithelium
(B) is usually at a high concentration in the lung of the premature infant
(C) causes the surface tension exerted on small alveoli to be less than that exerted on large alveoli
(D) increases surface tension
(E) is not formed by alveolar cells

392. The alveolar pressure is

(A) higher (less negative) than the intrapleural pressure during inspiration
(B) higher (less negative) than the intrapleural pressure during expiration
(C) higher (less negative) than the intrapleural pressure during inspiration and expiration
(D) lower (more negative) than the intrapleural pressure during expiration
(E) lower (more negative) than the intrapleural pressure during inspiration and expiration

393. In emphysema,

(A) a forced expiration causes a collapse of some of the bronchioles
(B) a forced expiration is essential for a normal tidal volume
(C) there is a decreased resistance to air flow on inspiration
(D) there is an increased resistance to air flow on inspiration
(E) there is a tendency for alveoli to collapse

394. In emphysema, there is an increased tendency for bronchioles to collapse during a forced expiration. This is due to

(A) a decrease in the elasticity of the lungs
(B) a loss of collaginous tissue from the lungs
(C) a failure of bronchiolar chondroblasts
(D) excessive tone in the bronchiolar smooth muscle
(E) hypersensitivity to epinephrine

395. Which of the following statements best characterizes the pattern of ventilation in the lungs during quiet breathing?

(A) surfactant keeps each region of the lung equally distended and ventilated
(B) gravity in the erect individual keeps the base of the lung (inferior portion) more poorly expanded and ventilated than the apex
(C) gravity in the erect individual keeps the base of the lung (inferior portion) more poorly expanded and better ventilated than the apex
(D) gravity in the erect individual keeps the base of the lung (inferior portion) more expanded and ventilated than the apex
(E) gravity in the erect individual keeps the base of the lung (inferior portion) more expanded and less ventilated than the apex

396. Which of the following factors increase respiratory minute work?

(A) airway constriction
(B) increased tidal volume
(C) A and B
(D) increased compliance of the lungs
(E) decreased density of the inspired gas

397. What is the most efficient method for the asthmatic to use during respiration?

(A) hyperventilation
(B) higher respiratory rate and lower tidal volume than the healthy subject
(C) lower respiratory rate and higher tidal volume than the healthy subject
(D) a respiratory rate and volume comparable to that of a healthy person
(E) a respiratory rate and volume comparable to that for a person with a reduced lung compliance

398. A mountain climber has been resting for an hour near the summit of Mount Everest (barometric pressure, 247 mm Hg). Under these circumstances

(A) at the end of inspiration, the air in her bronchioles will have a PO_2 of less than 45 mm Hg
(B) at the end of inspiration, the air in her bronchioles will have a PH_2O of less than what is found at sea level (ie, less than 35 mm Hg)
(C) A and B
(D) her arterial blood will have a lower pH than what is found at sea level
(E) all of the above

399. The respiratory exchange ratio (R)

(A) decreases during strenuous running
(B) decreases during metabolic acidosis
(C) decreases during both of the above
(D) increases during hyperventilation
(E) increases in individuals who shift from a high-carbohydrate diet to a high-fat diet

400. Which one of the following functions is not performed by the lungs?

(A) preventing emboli from moving through the lungs to the aorta
(B) fibrinolysis
(C) activation of angiotensin I to angiotensin II
(D) removal from the blood of epinephrine, vasopressin, and oxytocin
(E) removal from the blood of adenine nucleotides, serotonin, and norepinephrine

Answers and Explanations

373. (A)

374. (B) The contraction of the diaphragm decreases intrathoracic pressure and increases intra-abdominal pressure. It is primarily elastic recoil that is responsible for expiration in a subject at rest. Elevation of the larynx would occlude the glottis and prevent inspiration.

375. (E) Cartilage is found from the trachea through the small bronchi. It is not normally found below the bronchi (ie, in the bronchioles and alveoli). Inspired dry air at –10°C has reached body temperature and 100% humidity by the time it leaves the nasal cavity. Mucus is being continuously carried by cilia from the nasal cavity and tracheobronchial tree toward the hypopharynx, where it is swallowed. Approximately 432 g of mucus are produced by the respiratory epithelium per hour [$(5.4\ g/m^2/hr) \times (80\ m^2)$].

376. (A) The external intercostal muscles serve to elevate the ribs and stabilize the intercostal space during inspiration. Their paralysis, in this case, is probably due to the muscle relaxant. The sucking inward of the soft tissue is due to the negative intrathoracic pressure during inspiration. Abdominal breathing during anesthetization is usually due to a loss of abdominal muscle tone, which produces a pronounced distention of the abdomen during inspiration.

377. (A) The scalene and sternomastoid muscles elevate the rib cage, and the trapezius muscle stabilizes the head so that contraction of the sternomastoid muscle does not move the head and thus dampen its action on the rib cage. The transversus abdominis muscle compresses the abdominal contents and pushes the diaphragm cephalad.

378. (C) In aging, there is a loss of the lungs' elastic pull on the thorax and a resultant increase in residual volume. At age 45, the residual volume averages 1.48 L, at age 65 it is 1.72 L, and at age 75 1.92 L. The residual volume is the volume of air left in the lungs after a maximal expiration. Atelectasis is a collapse of the alveoli.

379. (C) Vital capacity = inspiratory reserve volume + resting tidal volume + expiratory reserve volume. Dead air space = tidal volume – the volume of fresh air that enters the respiratory portion of the respiratory system during an inspiration. Inspiratory reserve volume = (vital capacity) – (expiratory reserve volume + resting tidal volume).

380. (C)

$$\text{Pulmonary ventilation} = \text{pulmonary ventilation at rest} \times (O_2 \text{ consumption during exercise})/O_2 \text{ consumption before exercise}$$
$$= (5\ L/min) \times 3\ L\ of\ O_2/min/0.25\ L\ of\ O_2/min) = 60\ L/min$$

Under these circumstances, the arterial PO_2 would be expected to stay fairly constant during exercise, and, therefore, there would be a direct relationship between the increase in O_2 consumption and the increase in pulmonary ventilation. Changes in cardiac output are not needed for this calculation.

381. (B)

$$\text{Dead space volume} = [(\text{alveolar } PCO_2 - \text{expired } PCO_2)/\text{alveolar } PCO_2] \times \text{tidal volume}$$
$$= (40 \text{ mm Hg} - 30 \text{ mm Hg})/40 \text{ mm Hg} \times 500 \text{ mL}$$
$$= (10/40) \times 500 \text{ mL} = 125 \text{ mL}$$

In the above equation, you are assuming that the PCO_2 of the inspired air is approximately 0 mm Hg. Since the dead air space in a healthy individual with a tidal volume of 500 mL varies from 100 to 200 mL, it would seem that the tube through which this subject breathes contributes little to his dead space.

382. (B)

$$\text{Dead space} = (150 \text{ mL} \times 100)/3000 \text{ mL}$$
$$= 5\% \text{ of functional residual capacity}$$

The anatomic dead space is larger at the total lung capacity because the more negative intrapleural pressure in this state causes dilation of the bronchi and bronchioles.

$$\text{Dead space} = (180 \text{ mL} \times 100)/6000 \text{ mL}$$
$$= 3\% \text{ of total lung capacity}$$

The physiologic dead space may be considerably greater than the anatomic dead space in certain pathologic conditions.

383. (C) A decrease in alveolar ventilation causes a decrease in the arterial PO_2. It will be associated with no change in respiratory minute volume when there is an increased "dead space minute volume" (= dead space × respiratory rate). See the formula for alveolar ventilation in Answer 379. Because respiratory minute volume = tidal volume × respiratory rate, there cannot be a constant minute volume associated with a decrease of both tidal volume and rate.

384. (E) In this study, the first gas inspired into the alveoli was from the dead space and therefore contained the highest concentration of N_2. Most of it moved into the upper portions of the lungs. During phase M, many of the airways in the lower portions have closed because of their low transmural pressures. Therefore, most of the last expired air comes from these N_2-rich upper lungs. They are also responsible for the slight slope seen in phase L. The gas at J is the first gas expired and the only gas containing no N_2. It is only from the dead space. K is the second phase of expiration. The gas is a mixture of dead space gas (0% N_2) and alveolar gas. L is the third phase and represents alveolar gas.

385. (C) The compliances (C) are represented by the slopes of lines F–G (subject X) and F–H (subject Y): slope = (L)/(cm H_2O) = C. The height of point G (subject X) and H (subject Y) equal the tidal volumes.

386. (E) There will be a lower intrathoracic pressure on the normal side that will tend to pull the mediastinum toward that side. A pneumothorax may result from a ruptured alveolus, an incision through the chest wall, or an injection of air through the chest wall. When it occurs, the lungs, owing to their elastic characteristics, recoil and the residual volume is reduced. The elastic characteristics of the lungs are such that at the end of either a resting or maximal expiration they are pulling the chest wall inward. Usually, pneumothorax causes the chest wall to obtain a size similar to that found at the end of a normal resting inspiration. The presence of air in the intrapleural space would dampen the inward pull of the lung on the chest wall and tend to make the wall expand rather than collapse. The intrapleural pressure would move away from its subatmospheric values of between –2 and –6 mm Hg toward atmospheric pressure. The pneumothorax would cause a dampening of the pressure changes during inspiration as a result of the presence of a distensible fluid (ie, air) in the intrapleural space.

387. **(B)** These two compliances are parallel. Therefore, the total compliance (Ct) is:

$$1/Ct = 1/c1 + 1/c2 = 1/0.2 + 1/0.2 = 2/0.2 = 1/0.1$$
$$Ct = 0.1 \text{ L/cm of } H_2O$$

388. **(D)** In aging there is an increased residual volume which causes the increased compliance and an infiltration of the lung with collagen which causes the decreased specific compliance. Obesity and pregnancy decrease compliance by interfering with an increase in the cephalad–caudad dimension of the thorax. In diffuse alveolar fibrosis, compliance may change from a normal value of 165 mL/cm of H_2O to 10 mL/cm of H_2O. In alveolar edema there is a decreased residual volume and decreased compliance. Loss of surfactant production causes an increased surface tension at the air–liquid interface of the alveoli and a decreased compliance.

389. **(D)** Passing air into the mouth with the external nares open will not expand the lung. The pressure needed to open the collapsed alveoli is greater than that needed to distend the lung once the alveoli are patent. The absence of surfactant makes the lung less compliant.

390. **(D)** At 26 weeks of gestation, the lecithin/sphingomyelin ratio is less than 1, but lecithin production has begun, and the lung is beginning to produce the dipalmitoyl lecithin-containing substance, surfactant. At 35 weeks, there is a rapid jump in lecithin production, but sphingomyelin concentration stays constant. As the ratio increases from less than 1 to greater than 2, the incidence of death due to respiratory distress syndrome (RDS) or hyaline membrane disease decreases. At a ratio of 1, there is a survival of about 10%. At a ratio greater than 2, there are no deaths due to RDS.

391. **(C)** Surfactant is a detergent-like substance that lowers surface tension. As the alveolus enlarges, the concentration of surfactant at the surface becomes less, and the surface tension therefore goes up. It is localized at the air–liquid interface. Its production begins late in fetal life, and the premature infant is frequently deficient in surfactant. This deficiency causes the atelectasis seen in hyaline membrane disease of the newborn. Surfactant decreases surface tension. It is formed by cells lining the alveoli.

392. **(C)**

393. **(A)**

394. **(A)** In a normal subject, the ability of the lungs to recoil during expiration helps to keep a higher pressure in the lumen of the bronchiole than there is in the peribronchiolar space. In emphysema, the loss of lung elasticity causes an increase in the residual volume of the lungs and bronchiolar occlusion during a forced expiration. Bronchioles do not contain cartilage, and cartilage-forming cells do not play a role in the patency of bronchioles. Excessive tone in the bronchiolar smooth muscle is more characteristic of allergic reactions than emphysema. Epinephrine produces a decrease in bronchiolar muscle tone.

395. **(C)** The intrapleural pressure at the base of the lung is less negative, and, therefore, lung volume in this region is also less than found in the upper portions of the lungs. The smaller volume of each inferior alveolus will mean that it will distend more than the larger cephalad alveoli during inspiration and, therefore, be better ventilated. Surfactant does serve to reduce overdistention, but there is good evidence for uneven ventilation of the alveoli.

396. **(C)** Both airway constriction and increased tidal volume increase respiratory minute work. Airway constriction increases airway resistance and therefore minute work. A decrease in compliance increases the respiratory minute work. An increase in density increases the respiratory minute work.

397. **(C)** The total work performed during inspiration equals (1) the work used to overcome elastic forces plus (2) the work used to overcome resistance. The asthmatic can maintain a normal alveolar ventilation and keep his efficiency optimal under the circumstances by increasing tidal volume and decreasing respiratory rate. During quiet breathing, 1.5% of the O_2 consumption occurs in the respiratory muscles. A sevenfold increase in ventilation causes 3% of the O_2 consumption to occur in respiratory muscles. This is the most efficient ventilation for an individual with a restrictive ventilatory insufficiency (lung fibrosis, obesity, etc.).

398. **(A)** In choice (B), the air should differ from the air outside the body in that it is at body temperature and has a PH_2O of 47 mm Hg (ie, at a humidity of 100%):

$$Po_2 = (247 - 47 \text{ mm Hg})\ (0.20) = 40 \text{ mm Hg}$$

The hypoxia that occurs at this altitude should cause a greater ventilation than would be found at sea level and therefore a lower arterial CO_2 and H^+ concentration in the arterial blood.

399. **(D)** R = (CO_2 expired)/(O_2 take-up by lungs and blood). It is increased by increases in ventilation and increases in respiratory quotient (RQ). During exercise, metabolic acidosis, and hyperventilation, there is an increased ventilation which produces a depletion of CO_2 stores without a comparable movement of O_2 into the body. Under steady-state conditions, the body's R and RQ are equal. Carbohydrate catabolism moves the RQ toward 1, and fat catabolism moves it toward 0.7.

400. **(D)** The lungs remove none of the substances listed in (D).

CHAPTER 17

Diffusion and Perfusion

Questions

DIRECTIONS (Questions 401 through 416): Each of the numbered items or incomplete statements in this section is followed by answers or by completions of the statement. Select the ONE lettered answer or completion that is BEST in each case.

401. The structures (or barriers) through which O_2 must diffuse in passing from the surfactant containing liquid in the alveolar lumen to hemoglobin are in the following order:

(A) alveolar membrane, basement membrane, plasma, capillary endothelium, erythrocyte membrane
(B) alveolar membrane, basement membrane, capillary endothelium, plasma, erythrocyte membrane
(C) alveolar membrane, plasma, basement membrane, capillary endothelium, erythrocyte membrane
(D) basement membrane, alveolar membrane, plasma, capillary endothelium, erythrocyte membrane
(E) capillary endothelium, alveolar membrane, plasma, capillary endothelium, erythrocyte membrane

402. All of the following are factors that determine the quantity of gas that will diffuse through a barrier EXCEPT

(A) surface area available for diffusion: A
(B) thickness of barrier: T
(C) molecular weight of diffusing particle: M
(D) viscosity of the medium: Eta
(E) driving pressure: P1 – P2

403. A series of gas mixtures is inhaled by a healthy subject. Which of the following gases would diffuse most slowly from the lungs into the blood?

(A) CO_2 at a P_{CO_2} of 60 mm Hg
(B) CO at a P_{CO} of 0.5 mm Hg
(C) O_2 at a P_{O_2} of 130 mm Hg
(D) O_2 at a P_{O_2} of 150 mm Hg
(E) nitrous oxide at a P_{N_2O} of 0.3 mm Hg

404. If an individual who is scuba diving 66 feet below the surface of the sea is exposed to a barometric pressure of 3 atmosphere, what will be the P_{N_2} in the inspired tracheal air of an individual who is 132 feet below the surface of the sea and is breathing air from his tank?

(A) less than 2500 mm Hg
(B) 2600 mm Hg
(C) 2800 mm Hg
(D) 3000 mm Hg
(E) over 3100 mm Hg

405. If an individual who is breathing air has been scuba diving for 10 minutes at a depth of 132 feet and rapidly rises to the surface, she will generally experience no ill effects. If an individual has been breathing air for 540 minutes at a depth of 90 feet, it is recommended that she go through a period of 720 minutes of decompression. If she does not, she may experience symptoms of the bends, such as

(A) unconsciousness
(B) pain in the extremities
(C) substernal distress
(D) spastic paralysis
(E) all of the above

406. Under what circumstances, in the healthy subject, will the Po_2 of the blood leaving a pulmonary capillary be lower than the Po_2 of the alveolus served by that capillary?

(A) breathing air with a high Po_2
(B) performing an exercise in which the cardiac output is tripled
(C) breathing air with a high Pco_2
(D) all of the above
(E) none of the above

407. Which of the following gases in the healthy subject is diffusion limited?

(A) CO_2
(B) CO
(C) A and B
(D) O_2
(E) nitrous oxide

408. A subject inspires a mixture of gases containing CO and holds his breath for 10 seconds. It is calculated that during those 10 seconds the alveolar Pco is 0.5 mm Hg and the CO uptake is 25 mL/min. What is the diffusing capacity for CO?

(A) 5 mL/min/mm Hg
(B) 15 mL/min/mm Hg
(C) 50 mL/min/mm Hg
(D) 150 mL/min/mm Hg
(E) 500 mL/min/mm Hg

409. A patient suffered from pulmonary congestion as a result of left ventricular failure. Which of the following is MOST consistent with the finding of a moderate diffusion block in this patient?

(A) lowered pulmonary vein pressure
(B) elevated Pco_2 in systemic arteries
(C) elevated Po_2 in systemic arteries
(D) lowered Po_2 in systemic arteries
(E) elevated Pco_2 and Po_2 in systemic arteries

410. A patient is noted to be cyanotic. You suspect that she has either (1) a reduced diffusing capacity or (2) a true right-to-left shunt due to a tetralogy of Fallot. Which of the following features distinguishes these two types of cyanosis?

(A) in (1) only, there will be an abnormally low arterial Po_2
(B) in (1) only, there will be an abnormally low alveolar Pco_2
(C) in (1) only, the patient will remain cyanotic when breathing 100% O_2
(D) in (2) only, there will be an abnormally low arterial Po_2
(E) in (2) only, the patient will remain cyanotic when breathing 100% O_2

411. In the erect subject, the perfusion of the lung via the pulmonary vessels is

(A) uniform throughout the lung
(B) increased during positive pressure artificial respiration if the alveolar pressure is kept above atmospheric pressure
(C) A and B
(D) greatest in the superior part of the lung
(E) greatest in the inferior part of the lung

412. In the healthy, erect adult, the arterial blood has a Po_2 that is about 4 mm Hg lower than that found in the mixed alveolar air. What is the mechanism(s) responsible for this?

(A) some of the alveoli with a low alveolar Po_2 receive a greater perfusion with blood than alveoli with a high Po_2

(B) the thebesian veins empty venous blood into the left ventricle
(C) bronchial vessels empty venous blood into the pulmonary vein
(D) all of the above contribute to the lower arterial PO_2
(E) the diffusion characteristics of the lung

413. Which of the following statements best characterizes the alveoli of the healthy lung in an erect subject? The alveoli

(A) have their highest ventilation/blood perfusion ratio in the superior part of the lung
(B) have their highest ventilation/blood perfusion ratio in the inferior part of the lung
(C) all have a ventilation/blood perfusion ratio of 1
(D) in the inferior part of the lung usually have a lower ventilation than those in the superior part of the lung
(E) B and D

414. What happens when the ventilation/perfusion ratio of a lung unit decreases? The alveoli in that unit develop a

(A) higher PO_2
(B) lower PN_2
(C) higher PO_2 and lower PCO_2
(D) higher PCO_2
(E) higher PN_2 and higher PO_2

415. A number of patients with an exaggerated ventilation/perfusion inequality have a lowered arterial PO_2 but a normal arterial PCO_2. Why is this?

(A) CO_2 diffuses more rapidly than O_2
(B) CO_2 is more soluble in the blood than O_2
(C) A and B
(D) the O_2 and CO_2 dissociation curves have different shapes
(E) there is a greater pressure head for CO_2 than O_2 between the alveoli and the pulmonary capillary blood

416. The following data were obtained from a patient:

Arterial PO_2	100 mm Hg
Right atrial PO_2	40 mm Hg
Diffusing capacity for O_2	30 mL/min/mm Hg
O_2 consumption	240 mL/min

What is the patient's average PO_2 difference between his alveolar gas and his alveolar capillary blood?

(A) 8 mm Hg
(B) 20 mm Hg
(C) 40 mm Hg
(D) 50 mm Hg
(E) 60 mm Hg

Answers and Explanations

401. (B) Surfactant containing liquid → alveolar membrane → basement membrane → capillary endothelium → plasma → erythrocyte membrane.

402. (D) Viscosity is not involved in determining diffusion rate. Solubility of the particle in the medium (S) is also a factor.

403. (B) CO is 0.047 times as soluble in body fluids as N_2O.

404. (D)

$$\text{Barometric pressure at 132 ft} = (1 \text{ atm}) + (132 \text{ ft}/33 \text{ ft per atm}) = 5 \text{ atm}$$

$$P_{N_2} = (0.8) \times [(5 \text{ atm} \times 760 \text{ mm Hg/atm}) - 47 \text{ mm Hg}] = 3002 \text{ mm Hg}$$

Note that in this solution (1) an allowance is made for the dilution of the inspired air with water vapor (P_{H_2O} = 47 mm Hg), (2) the assumption is made that 80% of the air in the tank is N_2, and (3) the conclusion is drawn that each 33 feet below the surface of the sea adds 1 atmosphere of pressure to the subject and the air he breathes.

405. (E) Development of the bends involves two variables: (1) the P_{N_2} and (2) the duration of exposure to that P_{N_2}. The subject who is most susceptible to the bends has a longer exposure to the elevated P_{N_2}. This longer exposure provides the time necessary for N_2, which diffuses slowly, to move into the perivascular tissues. It necessitates a longer period of decompression for N_2 to diffuse out of the body. It has been estimated that it takes 1 hour for a subject breathing 100% O_2 to lose 90% of her N_2, and 3 hours for her to lose 95% of her N_2. If decompression is too rapid, N_2 comes out of solution and forms N_2 bubbles in the tissues and sometimes in the blood.

406. (E) Under resting conditions, the blood stays in the pulmonary capillary for about 0.75 second and acquires the alveolar P_{O_2} during the first 0.3 second. In exercise, the cardiac output increases and the blood passes through the pulmonary capillaries more rapidly. When the cardiac output is increased threefold or less, there still is an equilibrium reached between the blood P_{O_2} and the alveolar P_{O_2} before the blood leaves the lungs. In other words, in the lung of the healthy subject, oxygen transfer is seldom if ever diffusion-limited.

407. (B) The quantity of CO that is carried from the lungs does not increase with increases in pulmonary blood flow. In other words, its transfer is not perfusion-limited but is diffusion-limited. CO_2, O_2, and nitrous oxide transfer in the healthy subject, on the other hand, is not diffusion-limited but is perfusion-limited.

408. (C)

$$D_{CO} = I_{CO}/(P_1CO - P_2CO) = 25 \text{ mL/min}/(0.5 - 0) \text{ (mm Hg)} = 50 \text{ mL/min/mm Hg}$$

In this procedure, it is assumed that P_2CO (CO pressure in pulmonary capillary blood) is 0.

409. (D) O_2 diffuses through barriers about one twentieth as readily as CO_2. Characteristic of a moderate diffusion block, therefore, is a hypoxia not associated with hypercapnea. An increased pressure in the pulmonary system is causing pulmonary edema and, hence, is causing the diffusion block. CO_2 diffuses through barriers much more readily than does O_2. In fact, the lowered arterial PO_2 produced in this condition can cause a reflex increase in alveolar ventilation and, thereby, a decrease in alveolar and arterial PCO_2. On the other hand, a severe diffusion block can cause an elevated arterial PCO_2.

410. (E) Breathing 100% O_2 does not appreciably increase the O_2 content of the blood leaving the lung of a normal patient or one with a tetralogy of Fallot. The reason for this is that in a normal individual or a patient with a tetralogy, most of the O_2 is carried in combination with hemoglobin, and the hemoglobin leaving the lung is over 95% saturated with O_2 when this individual is breathing room air. On the other hand, one can overcome a diffusion block by increasing the PO_2 of the inspired air. A low arterial PO_2 is characteristic of both conditions. The alveolar PCO_2 may be the same in both conditions. A reduced diffusing capacity for O_2 may be associated with a normal diffusing capacity for CO_2. In diffusion blocks, it is possible to deliver normal quantities of O_2 to the blood by increasing the PO_2 of the inspired air.

411. (E) In the erect individual, the weight of the blood tends to make the arterial, capillary, and venous pressure in the most inferior part of the lung about 23 mm Hg higher than in the most superior part of the lung. This contributes to a smaller resistance to flow in the inferior part of the lung than in the superior part. An increased alveolar pressure tends to occlude pulmonary vessels and, therefore, decrease the perfusion of the lung.

412. (D) The alveoli in the superior part of the lung may have a PO_2 of 130 mm Hg, while those in the inferior part may have a PO_2 of 90 mm Hg. The inferior alveoli have a greater blood flow. O_2 transfer is not diffusion-limited in the healthy adult.

413. (A) As one goes from the superior to the inferior part of the lung, there is an increase in ventilation and blood perfusion but a decrease in ventilation/blood perfusion ratio. In other words, the ventilation increases less than the blood perfusion.

414. (D) There will be a higher PCO_2 and a lower PO_2 in the alveoli.

415. (D) Since the CO_2-dissociation curve is not "S"-shaped (ie, not plateaued), but is closer to a straight line, alveoli with a high ventilation/perfusion ratio (low PCO_2) will compensate for alveoli with a low ventilation/perfusion ratio (high PCO_2). This is not true for O_2. Alveoli with a PO_2 of 90 and 180 mm Hg add almost the same amount of O_2 to the blood. In the healthy subject, the blood, before it leaves the pulmonary capillary, has come into equilibrium with the O_2 and CO_2 in the alveoli it serves (ie, CO_2 and O_2 are not diffusion-limited in this case). Therefore, the speed of diffusion is not an important factor. The pressure head for CO_2 is about 6 mm Hg and that for O_2 is about 60 mm Hg.

416. (A)

$$\begin{aligned}\text{Driving pressure} &= \text{flow/diffusing capacity}\\ &= 240\ \text{mL/min}/30\ \text{mL/min/mm Hg}\\ &= 8\ \text{mm Hg}\end{aligned}$$

CHAPTER 18

Control of Respiration

Questions

DIRECTIONS (Questions 417 through 436): Each of the numbered items or incomplete statements in this section is followed by answers or by completions of the statement. Select the ONE lettered answer or completion that is BEST in each case.

417. Which of the following procedures will cause an immediate cessation of respiration?

(A) transection of the cord at C6
(B) transection of the cord at C2
(C) transection between the medulla and pons
(D) all of the above procedures cause apnea
(E) none of the above procedures causes apnea

418. Which of the following statements is correct? The medullary inspiratory center

(A) responds to mild increases in P_{CO_2} and acidity in the cerebrospinal fluid (CSF) of the fourth ventricle of the brain by increasing the number of inspirations per minute
(B) responds to mild increases in P_{CO_2} and acidity in the CSF of the fourth ventricle of the brain by increasing the depth of inspiration
(C) responds as stated in both A and B
(D) when stimulated, stimulates the medullary expiratory center
(E) none are correct

419. The healthy heart has its pacemaker in the sinoatrial node. The stimuli for the contraction of the diaphragm originate

(A) in the floor of the fourth ventricle of the brain
(B) in the pneumotaxic center
(C) in the apneustic center
(D) in chemoreceptors
(E) in stretch receptors

420. A cat has its ninth and tenth cranial nerves severed bilaterally, and a section is made in the pons just above the apneustic center. How do these procedures modify the respiratory pattern? They cause

(A) decerebrate rigidity
(B) long inspiratory gasps
(C) A and B
(D) apnea
(E) eupnea

421. Which of the following statements is correct? The pneumotaxic center

(A) is in the midbrain
(B) inhibits inspiratory activity
(C) contains the major central chemoreceptor area
(D) causes long inspiratory gasps when separated from the more superior parts of the brain
(E) causes long expiratory gasps when separated from the more superior parts of the brain

422. Which of the following structures is MOST important in increasing respiratory minute volume in response to a small increase in the PCO_2 of the body fluids?

(A) pulmonary chemoreceptors
(B) venous chemoreceptors
(C) lung receptors
(D) the hypothalamus
(E) medullary chemoreceptors

423. The peripheral chemoreceptors are most important because they respond to

(A) decreases in PO_2 in the venous blood
(B) decreases in PO_2 in the arterial blood
(C) decreases in PO_2 in the cerebrospinal fluid
(D) increases in PO_2 in the venous blood
(E) increases in PO_2 in the arterial blood

424. Chemoreceptors in the carotid and aortic bodies send impulses via the ninth and tenth cranial nerves to the respiratory centers. Which of the following BEST characterizes their function? They

(A) send increasing frequencies of impulses up their nerves as the PO_2 of arterial blood increases
(B) produce a more rapid increase in ventilation in response to an increased arterial PCO_2 than do the central chemoreceptors
(C) are less sensitive to hypoxia than the central chemoreceptors
(D) are least important in the control of respiration during sleep and barbiturate depression
(E) affect only respiratory rate

425. The peripheral chemoreceptors produce a more pronounced increase in ventilation in response to

(A) a decrease in arterial PO_2 from 150 to 90 mm Hg than from 70 to 40 mm Hg under the usual resting conditions
(B) a change in arterial PO_2 from 100 to 80 mm Hg at a PCO_2 of 48 mm Hg than at a PCO_2 of 40 mm Hg
(C) A and B
(D) a 30% reduction in the O_2 content of the arterial blood, as in anemia, than to a 30% reduction in arterial PO_2
(E) a change in pH from 7.4 to 7.3 than cyanide poisoning

426. If the ninth and tenth cranial nerves are blocked in the neck, the subject will no longer respond to

(A) hypercapnea by causing an increased respiratory minute volume
(B) alkalosis by causing an increased respiratory minute volume
(C) hypoxia by causing an increased respiratory minute volume
(D) hypercapnea or acidity by causing an increased respiratory minute volume
(E) acidity or hypoxia by causing an increased respiratory minute volume

427. Which of the following conditions usually causes an increased frequency of impulses in afferent neurons from the carotid bodies?

(A) CO poisoning
(B) anemia
(C) A and B
(D) cyanide poisoning
(E) a 50% reduction in carotid body blood flow

428. The carotid bodies differ from the

(A) aortic bodies in that the carotid bodies are sensitive to a decreased arterial PO_2 and the aortic bodies are not
(B) respiratory center in the medulla in that the carotid bodies have an opposite response to a decreased arterial PO_2
(C) aortic bodies and respiratory center as stated above
(D) aortic bodies in that the carotid bodies are sensitive to changes in PCO_2
(E) respiratory center in that the respiratory center is sensitive to arterial PCO_2

429. It has been noted in breath-holding experiments that lung distention increases the tolerance of the body for CO_2 and hypoxia. What is the mechanism of action?

(A) stimulation of stretch receptors in the bronchi and lung parenchyma
(B) stimulation of sensory neurons in the vagus nerve
(C) inhibition of respiratory centers
(D) A and B
(E) A, B, and C

430. Which of the following statements is correct? Further inspiratory activity is inhibited during inspiration

(A) by slowly adapting stretch receptors in the adult lung when the inspiratory depth exceeds 1 liter
(B) by rapidly adapting stretch receptors in the adult lung when the inspiratory depth exceeds 1 liter
(C) by the action of inhaled irritants on the respiratory mucosa
(D) A and C
(E) none are correct

431. A healthy subject responds to running at a moderate rate with an increased ventilation caused by

(A) a decreased venous P_{O_2}
(B) a decreased arterial P_{O_2}
(C) proprioceptive impulses from moving limbs
(D) an increased arterial pH
(E) an increased arterial P_{CO_2}

432. A patient is brought to the emergency room suffering from an overdose of a barbiturate. He exhibits hypoventilation due to respiratory center depression. He is given 100% O_2 and his respiratory minute volume decreases markedly, but his mixed venous plasma P_{O_2} rises to 130 mm Hg. The patient probably

(A) is now well oxygenated and needs no additional treatment
(B) should be switched to 95% O_2 + 5% CO_2
(C) should receive a vasoconstrictor agent
(D) should be treated for systemic acidosis
(E) should be treated for systemic alkalosis

433. What useful changes occur when one becomes acclimatized to a high altitude?

(A) hyperventilation
(B) polycythemia
(C) increased number of systemic capillaries
(D) O_2 dissociation curve shifts to the right
(E) all of the above

434. A decrease in plasma pH

(A) causes an increase in ventilation through the stimulation of the carotid bodies
(B) is frequently associated with a decrease in arterial P_{CO_2} and an increase in ventilation in metabolic acidosis
(C) A and B
(D) may not cause as great a decrease in the pH of the cerebrospinal fluid because the blood–brain barrier and blood–cerebrospinal fluid barrier are not freely permeable to H^+
(E) all of the above

435. Factors that tend to increase the pulmonary arterial pressure in individuals who have lived at altitudes of 12,000 feet for a week include

(A) an increased hematocrit
(B) a low alveolar P_{O_2}
(C) A and B
(D) acidosis
(E) A and D

436. A moderate hypoxia differs from a moderate asphyxia in that in hypoxia, the direct

(A) stimulation of the central chemoreceptors is a more important mechanism for increasing ventilation
(B) stimulation of the aortic and carotid sinus is a more important mechanism for increasing ventilation
(C) stimulation of the central and peripheral chemoreceptors is an equally important mechanism for increasing ventilation
(D) stimulation of the aortic and carotid bodies is a more important mechanism for increasing ventilation
(E) inhibition of the peripheral chemoreceptors is a more important mechanism for increasing ventilation

Answers and Explanations

417. **(B)** Somatic efferent neurons originating in C3, C4, and C5 pass down the right and left phrenic nerves to initiate the contraction of the diaphragm. They are not dependent upon impulses originating below C5 for their stimulation but are dependent upon impulses traveling down the cord from the medulla.

418. **(C)** The inspiratory and expiratory centers are mutually inhibitory; that is, the stimulation of one causes the inhibition of the other.

419. **(A)** There is an area in the medulla that sends out volleys of stimuli to the neurons of the phrenic nerves. This area continues to fire periodically when isolated from most of the rest of the body. When this area is depressed, it may become dependent upon impinging stimuli from peripheral chemoreceptors and elsewhere for continued function.

420. **(C)**

421. **(B)** The pneumotaxic center is superior to the apneustic center in the pons. The major central chemoreceptor area is in the medulla.

422. **(E)** If a resting subject inhales 5% CO_2, her ventilation will increase three- to fourfold. This great sensitivity to changes in P_{CO_2} is due to the central chemoreceptors.

423. **(B)** Changes in venous P_{O_2} are important if they produce changes in the arterial P_{O_2}. The central chemoreceptors are more sensitive to changes in P_{CO_2} than the peripheral chemoreceptors. CO_2, unlike H^+, passes readily through the blood–brain barrier.

424. **(B)** Although peripheral receptors are less sensitive to changes in P_{CO_2}, it takes longer for an increased P_{CO_2} from exercise or breath-holding to reach the medulla than the arterial receptors. They are stimulated by decreases in P_{O_2}. Small decreases in P_{O_2} that stimulate the peripheral chemoreceptors have no known action on the central receptors. Large decreases in P_{O_2} that stimulate the peripheral receptors decrease ventilation if the peripheral receptors are decentralized. During sleep, when the reticular activating center is less active, and during barbiturate depression, the peripheral receptors become increasingly important in the maintenance of respiration. They affect respiratory rate and depth.

425. **(B)** The sensitivity of the receptors to decreases in P_{O_2} will depend upon the P_{CO_2}. Under resting conditions, decreases in arterial P_{O_2} from 500 to 100 mm Hg produce only small changes in ventilation, whereas decreases from 70 to 40 mm Hg can produce better than a doubling of ventilation. When a decreased O_2 content is so severe that there is a decreased P_{O_2} in carotid body blood, then there will be an increased ventilation. Cyanide poisoning, by preventing the peripheral chemoreceptors from utilizing O_2, causes them to increase ventilation. The change in pH mentioned above has less effect on these receptors.

426. **(C)** Hypercapnea acts directly on the respiratory centers as well as on the peripheral receptors to increase ventilation. When the peripheral chemoreceptors are decentralized, breathing 8% O_2 has no action on ventilation, but when they are functioning, they will produce a threefold increase in respiratory rate and a sevenfold increase in respiratory depth.

427. **(D)** Either cyanide or a low arterial P_{O_2} will stimulate the carotid bodies. Cyanide prevents the utilization of oxygen (histoxic hypoxia) and the low P_{O_2} (hypoxic hypoxia) exposes the chemoreceptors to a lower concentration of oxygen. Both (A) and (B) are examples of anemic hypoxia. In both cases, there is a reduction in the oxygen content of the blood, but not in the P_{O_2} of the blood passing through the carotid bodies. Arterial P_{O_2} is independent of oxyhemoglobin concentration. The P_{O_2} of the blood leaving the carotid bodies is also relatively independent of oxyhemoglobin concentration because of the high level of flow through the bodies. The high blood flow that passes through the carotid bodies (80 mL/min or 2000 mL/200 g of tissue) makes them insensitive to most reductions in flow, as well as reductions in oxyhemoglobin concentration.

428. **(C)** If the carotid bodies are decentralized, hypoxic hypoxia causes a depression of respiration. The peripheral and central chemoreceptors are similar in that they are stimulated by increases in P_{CO_2} or H^+. On the other hand, the carotid bodies differ from the other chemoreceptors in that they are more sensitive to increases in H^+ concentration.

429. **(E)** The stimulation of stretch receptors in the lung has been shown to both facilitate and inhibit inspiration. When lung distention produces an inhibition of inspiration, this is said to be the Hering–Breuer reflex.

430. **(D)** Stretch receptors were, at one time, considered to be important in adults during quiet breathing but are now generally believed to function in the adult only when the tidal volume exceeds 1 L (in exercise, for example). The rapidly adapting receptors facilitate further inspiration. Their importance in human beings is uncertain. Irritants in the upper and lower respiratory passages cause a reflex inhibition of both inspiration and expiration, as well as reflex bronchial constriction. Irritants can also bring about an increased inspiratory effort followed by a cough.

431. **(C)** Human beings characteristically respond to running with a decreased venous P_{O_2}, little or no change in arterial P_{O_2}, an increased facilitation of respiratory activity by sensory impulses from the appendages, a decreased arterial pH due, in part, to the increased lactic acid concentration in the blood, and a decreased arterial P_{CO_2}. The decreased venous P_{CO_2} apparently is of little importance, since the venous side seems to lack effective hypoxia-sensing elements. Apparently, the increased ventilation that occurs during running is sufficient to prevent an important decreased arterial P_{O_2} and an important increased arterial P_{CO_2}, but not sufficient to prevent a decreased arterial pH.

432. **(D)** The inadequate ventilation is causing respiratory acidosis. Barbiturates, morphine, and other narcotics depress the respiratory centers. This results in a hypoxic hypoxia, which acts through the peripheral chemoreceptors to maintain respiration if the depression is not excessive. By giving 100% O_2, you treat the hypoxia but in so doing remove some of the stimulation to the respiratory centers. The further reduction in ventilation that results exaggerates the existing hypercapnea and therefore increases respiratory acidosis. This condition should be treated by the administration of $NaHCO_3$. Although it is true that by switching a subject with a normal P_{CO_2} from breathing 100% O_2 to breathing 95% O_2 + 5% CO_2 his ventilation can be doubled or tripled, this principle does not hold in this case. The subject is hypercapnic because of his hypoventilation. Increasing his hypercapnea will have little effect on ventilation and will increase the associated acidosis. At a PCO_2 greater than 76 mm Hg, ventilation is decreased by the addition of CO_2 to the arterial blood. The high venous PO_2 is not a serious problem. Oxygen toxicity, if present, is best treated by lowering the alveolar PO_2.

433. **(E)** Adapations to high altitude include the following:

1. Hyperventilation: increased respiratory rate, arterial PO_2, and arterial pH; decreased arterial PCO_2
2. Polycythemia: increased hemoglobin concentration in the blood; increased O_2-carrying capacity of the blood; increased blood volume
3. Increased concentration of systemic capillaries; increased efficiency during exercise
4. O_2 dissociation curve shifts to the right: increased concentration of 2,3-diphosphoglycerate; hemoglobin releases more O_2 at a PO_2 of 40 mm Hg

434. **(E)**

435. **(C)** The increased hematocrit causes an increased blood viscosity. The low alveolar PO_2 causes pulmonary vasoconstriction. The increased ventilation at high altitudes causes alkalosis.

436. **(D)** In asphyxia, there is hypoxia plus hypercapnea. Hypercapnea stimulates both the central and peripheral (aortic and carotid bodies) chemoreceptors. Hypoxia stimulates the peripheral receptors to send more nerve impulses to the respiratory centers. Its direct action on the centers is depression.

PART VI

Urine Formation

CHAPTER 19

Filtration, Reabsorption, and Secretion

Questions

DIRECTIONS (Questions 437 through 468): Each of the numbered items or incomplete statements in this section is followed by answers or by completions of the statement. Select the ONE lettered answer or completion that is BEST in each case.

437. In the kidney, the glomerular capillary pressure of a healthy adult

(A) is lower than the pressure in the efferent renal arteriole
(B) decreases when the afferent renal arteriole constricts
(C) rises by 8% to 10% when the aortic pressure rises by 10%
(D) is usually less than the pressure in a patent capillary in the deltoid muscle
(E) all of the above

438. The glomerulus

(A) is impermeable to all molecules with a molecular weight (MW) over 5000
(B) contains no active transport systems ("pumps") that produce an important effect on the composition of the glomerular filtrate
(C) produces a filtrate with a lower concentration of amino acids than found in plasma
(D) produces a filtrate with a higher concentration of urea than found in plasma
(E) all of the above

439. The glomerular filtration rate

(A) is greater than 50% of the plasma flow to the glomeruli
(B) falls to approximately 25% of normal when the mean arterial pressure changes from 100 to 25 mm Hg
(C) is decreased by a decrease in plasma colloid osmotic pressure
(D) increases ipsilateral to a ureteral obstruction
(E) none of the above

440. Which of the following is correct with regard to glucose and amino acid?

(A) reabsorption is most marked in the distal convoluted tubule
(B) transport is primarily by active secretion into the tubular fluid
(C) transport from the lumen of the nephron depends on Na^+ transport
(D) transport is controlled by parathormone
(E) transport is blocked by aldosterone

441. In the distal convoluted tubule, the cells

(A) contain large quantities of carbonic anhydrase which they use in secretion of H^+
(B) can reabsorb Na^+ in exchange for H^+ secretion across the luminal membrane
(C) A and B
(D) reabsorb over 40% of the glomerular filtrate
(E) determine the final composition of the urine

442. A patient is treated with a drug that inhibits the action of carbonic anhydrase. Which symptom is he LEAST likely to have?

(A) an increased urinary excretion of K^+
(B) an increased urinary excretion of Na^+
(C) A and B
(D) a decreased plasma pH
(E) a decreased urine volume

443. The renal clearance of

(A) a substance is measured in mg/mL
(B) a substance is measured in mg/min
(C) sodium is decreased by the injection of aldosterone
(D) inulin, at a plasma concentration of 60 mg%, is lower than at a plasma concentration of 120 mg%
(E) para-aminohippurate (PAH) at a plasma concentration of 60 mg% is higher than at a plasma concentration of 120 mg%

444. If a substance has a transport maximum (Tm) for reabsorption, this means that

(A) reabsorption is only passive
(B) only a constant fraction of the substance will be reabsorbed
(C) A and B
(D) below a threshold level, all of the substance will be reabsorbed
(E) phlorhizin blocks reabsorption

445. The renal plasma flow can be estimated by using a formula for flow that is similar to the formula devised by Fick for the estimation of cardiac output. In this procedure,

(A) para-aminohippurate (PAH) is injected at a rate such that its tubular maximum will not be exceeded
(B) inulin is injected at a rate such that its tubular maximum will not be exceeded
(C) para-aminohippurate is injected at a rate such that its tubular maximum will be exceeded
(D) inulin is injected at a rate such that its tubular maximum will be exceeded
(E) urea is injected

446. During the infusion of PAH into a patient, the concentration of PAH in the cephalic vein stabilized at 0.02 mg/mL of plasma (= PPAH). At this time, the two kidneys were producing 1 mL of urine per minute (= VU), and the concentration of PAH in the urine was 16 mg/mL (= UPAH). What was the PAH clearance (= CPAH)? What was the effective renal plasma flow (= ERPF)?

(A) 700
(B) 800
(C) 900
(D) 1000
(E) 1200

447. In a healthy individual, what percentage of the effective renal plasma flow would you expect to pass into the glomerular capsule?

(A) less than 5%
(B) between 15% and 20%
(C) between 40% and 50%
(D) between 70% and 80%
(E) greater than 90%

448. Which of the following statements is most consistent with a filterable substance being actively reabsorbed from the renal tubular lumen?

(A) its renal clearance value is lower than that of inulin
(B) its renal clearance value is higher than that of inulin
(C) the ratio of its rate of urinary excretion/plasma concentration is the same as that for glucose
(D) the ratio of its rate of urinary excretion/plasma concentration is greater than that for glucose
(E) its concentration in the distal tubule is higher than that in plasma

449. Which of the following substances does not have a Tm value?

(A) albumin, arginine
(B) β-hydroxybutyrate
(C) glucose

(D) phosphate, sulfate
(E) urea

450. The ratio of the amount of inulin excreted per minute to its arterial plasma concentration in the resting subject will

(A) decrease in severe bilateral ureteral obstruction
(B) increase in response to a decrease in plasma colloid osmotic pressure
(C) change as noted in A and B
(D) be dependent on active transport mechanisms
(E) be positively correlated with the rate of inulin infusion

451. A substance that has a renal clearance 20 times that of inulin is probably

(A) only filtered at the glomeruli
(B) only secreted by the tubules
(C) filtered and secreted
(D) synthesized in the tubules and secreted
(E) filtered and reabsorbed

452. Which of the following is correct? Urea

(A) is secreted by the distal convoluted tubule
(B) has a clearance greater than that for inulin
(C) both A and B are correct
(D) is formed primarily in the ascending part of the loop of Henle and in the distal convoluted tubule
(E) clearance increases as the volume of urine excreted increases

453. Which of the following substances is the major source of urea?

(A) dietary purines
(B) dietary proteins
(C) dietary pyrimidines
(D) dietary phospholipids
(E) β-hydroxybutyric acid

454. Which of the following statements BEST characterizes substance f seen in Figure 19–1? Substance f in the nephron is

(A) secreted
(B) reabsorbed
(C) filtered
(D) filtered and reabsorbed
(E) filtered and secreted

455. Which of the following statements BEST characterizes substance h in Figure 19–1? Substance h in the nephron is

(A) filtered and actively secreted
(B) filtered and passively reabsorbed
(C) filtered, passively reabsorbed, and actively reabsorbed
(D) filtered and synthesized
(E) filtered, synthesized, and secreted

456. Which of the following statements concerning curve h in Figure 19–1 is correct? The plasma concentration at point

(A) i represents the transport maximum
(B) i represents the splay
(C) i represents the threshold
(D) j represents the splay
(E) j represents the threshold

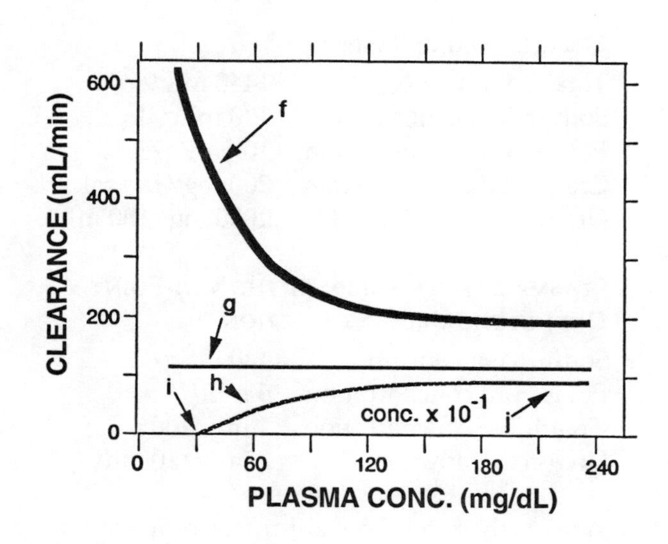

Figure 19–1. The clearances of substances f, g, and h were studied at different concentrations in the blood.

457. Toward the end of World War II, Karl Beyer and his associates noted that the injection of PAH decreased the excretion of penicillin in the urine. What would you suggest was its mechanism of action? The PAH

(A) competes with penicillin for a site on a carrier molecule in one of the reabsorptive mechanisms
(B) prevents active reabsorption
(C) either A or B could be correct
(D) increases filtration
(E) competes with penicillin for a site on a carrier molecule in one of the secretory mechanisms

458. In arthritis, there is the deposition of urate crystals in the joints. If you wished to treat arthritis by decreasing the concentration of circulating uric acid, you could

(A) keep your patient's caloric intake constant and carbohydrate intake low
(B) block active reabsorption of uric acid in the kidney
(C) A and B
(D) block active secretion of uric acid in the kidney
(E) A and D

459. The following data are obtained from a patient:

24-Hour Urine Sample

Total volume	1440 mL
Sodium concentration	120 mEq/L
Potassium concentration	100 mEq/L
Creatinine concentration	200 mg/100 mL
Urea concentration	2050 mg/100 mL

Plasma Sample Taken at the Mid-Point During the Urine Collection

Sodium concentration	140 mEq/L
Potassium concentration	5 mEq/L
Creatinine concentration	1 mg/100 mL
Urea concentration	25 mg/100 mL

What is the rate of potassium excretion?

(A) less than 0.2 mEq/min
(B) 0.2 mEq/min
(C) 0.3 mEq/min
(D) 0.4 mEq/min
(E) more than 0.4 mEq/min

460. On the basis of the data for the case in Question 459, the urea clearance is

(A) less than 78 mL/min
(B) 78 mL/min
(C) 80 mL/min
(D) 82 mL/min
(E) more than 82 mL/min

461. On the basis of the data for the case in Question 459, the rate of sodium reabsorption is

(A) less than 28 mEq/min
(B) 29 mEq/min
(C) 31 mEq/min
(D) 33 mEq/min
(E) more than 34 mEq/min

462. On the basis of the data from the case in Question 459, the fraction of the filtered urea that was excreted is

(A) 0.04
(B) 0.1
(C) 0.2
(D) 0.4
(E) 1.0

463. A mechanism in the kidney that produces marked increases in the tonicity of the peritubular liquid that surrounds the loop of Henle is called countercurrent multiplication. Which one of the following relationships is NOT part of this mechanism?

(A) the impermeability of the ascending limb prevents the diffusion of Na^+ through the cells that form the limb
(B) active transport of Na^+ into the peritubular space from the lumen of the descending limb
(C) active transport of Na^+ into the peritubular space from the lumen of the ascending limb

(D) diffusion of Na^+ into the descending limb
(E) hypertonicity of the fluid in the loop of Henle

464. The loop of Henle differs from the medullary collecting duct in that the loop of Henle

(A) in its descending limb, in the presence of ADH, is permeable to water and the collecting duct is not
(B) in its thick ascending limb actively transports ions that serve to make the medullary extracellular substance markedly hypertonic and the collecting duct does not
(C) in its thin segment contains, in the presence of ADH, a hypertonic solution and the collecting duct does not
(D) A, B, and C are correct
(E) A, B, and C are incorrect

465. Which of the following is correct? The juxtamedullary nephrons

(A) constitute over 40% of the nephrons in the kidney
(B) have a short loop of Henle
(C) lie solely in the renal medulla
(D) lie solely in the renal cortex
(E) are not well characterized by any of the above statements

466. During an experiment on a man, fluid from the last part of the proximal tubule was sampled after the administration of a drug. This sample had the following characteristics:

Inulin concentration	30% higher than that found in plasma
Na^+ concentration	150 mM/L

What is the most likely action of the drug? It

(A) decreased the glomerular filtration rate
(B) increased inulin secretion into the proximal tubule
(C) A and B
(D) decreased NaCl transport from the proximal tubule
(E) decreased urea transport in the proximal tubule

467. An osmotic diuresis resulting from hyperglycemia differs from water diuresis in that in the osmotic diuresis there is

(A) a substantial decrease in the reabsorption of water in the proximal tubule and the proximal part of the loop of Henle
(B) a decreased K^+ excretion
(C) an increased concentration of urea in the urinary bladder
(D) a decreased concentration of circulating antidiuretic hormone
(E) none of the above

468. Ammonia produced by the kidneys comes mainly from

(A) glycine
(B) glutamine
(C) leucine
(D) alanine
(E) B and D

DIRECTIONS (Questions 469 through 473): Each set of items in this section consists of a list of lettered options followed by several numbered words or phrases. For each numbered word or phrase, select the ONE lettered option that is most closely associated with it. Each lettered option may be selected once, more than once, or not at all.

(A) Bowman's capsule
(B) proximal tubule
(C) descending loop of Henle
(D) ascending loop of Henle
(E) distal tubule
(F) collection duct

469. Where does Tm-limited reabsorption of Na^+ occur?

470. Where does over 60% of the reabsorption of water occur?

471. Where is the luminal membrane that is least permeable to water in the presence of ADH?

472. What structure(s) contain(s) a hypertonic urine in the absence of ADH?

473. Where is the site of action of vasopressin?

Answers and Explanations

437. (B) The constriction of the afferent arteriole increases the resistance to flow into the capillary. The efferent arteriole carries blood from the glomerular capillary and therefore would be exposed to either a lower or the same pressure as in the capillary. At aortic pressures above 70 mm Hg, the afferent and efferent arterioles tend to keep the pressure in the glomerular capillaries constant. The pressure in the glomerular capillary is about 50 mm Hg and that in the deltoid capillary is about 30 mm Hg.

438. (B) Active transport is important in other parts of the nephron, but the major job of the glomerulus is to produce an ultrafiltrate of plasma. The glomeruli of the two kidneys produce approximately 125 mL of this filtrate per minute. Inulin has an MW of 5500 and has a concentration in the glomerular filtrate that is 98% of its concentration in plasma. Hemoglobin has an MW of 68,000 and has a concentration in the filtrate that is 3% of that in the plasma. The concentrations of amino acids, urea, glucose, ketone bodies, fatty acids, and dissolved electrolytes in the filtrate are the same as those in plasma. The main factor that keeps a small particle from filtration is protein binding. This occurs in the case of calcium, potassium, and a number of the hormones.

439. (E) The glomerular filtration rate (GFR) is normally about 125 mL/min and the plasma flow to the glomeruli is about 660 mL/min. It falls to 0 at arterial pressures less than 70 mm Hg. A decrease in plasma colloid osmotic pressure will cause either no change or an increased GFR. A ureteral obstruction will decrease the filtration pressure across the glomerulus and, therefore, decrease GFR.

440. (C) Glucose and Na^+ bind to a common carrier in a membrane lining the nephron. Parathormone controls Ca^{++} and phosphate reabsorption. Aldosterone facilitates Na^+ transport in the distal tubule.

441. (C) Carbonic anhydrase catalyzes the formation of H_2CO_3 from CO_2 and H_2O. Approximately 80% of the glomerular filtrate is reabsorbed in the proximal convoluted tubule. The collecting ducts also modify the composition of the urine.

442. (E) The increased excretion of Na^+ and K^+ will increase the urine volume. In the absence of carbonic anhydrase (CA) activity, there is less H^+ excreted and less HCO_3^- reabsorbed. Because Na^+ and K^+ reabsorption are paired with H^+ excretion and HCO_3^- reabsorption, Na^+ and K^+ reabsorption will be reduced (ie, Na^+ and K^+ excretion will be increased) when CA activity is reduced. Less excretion of H^+ causes more plasma H^+ (ie, a decreased pH).

443. (C) Aldosterone increases the reabsorption of Na^+ from the nephron and therefore decreases sodium clearance. Clearance is a measure of the mL of plasma cleared (ie, completely depleted) of a particular substance, x, per minute and can be calculated as follows:

$$Cx = \text{quantity of x excreted (mg/min)/}$$
$$\text{concentration of x in plasma (mg/mL)}$$

Inulin clearance is the same at a plasma concentration of 60 mg%. It is filtered but not secreted or reabsorbed. PAH clearance is less at a plasma concentration of 120 mg%. It is filtered and secreted.

444. (D) Phlorhizin inactivates the carrier system for (1) glucose, fructose, galactose, and xylose but does not affect the carriers for (2) sulfate and thiosulfate, nor for (3) arginine, lysine, ornithine, and cystine.

445. (A) PAH, when injected at a rate of less than 80 mg/min (= TmPAH), will have a renal vein concentration of almost zero and a renal artery concentration equal to that in an arm vein. This makes it easy to estimate the renal arteriovenous concentration difference for PAH:PAH concentration in arm vein – 0. Neither inulin (filtered) nor urea (filtered and reabsorbed) is completely removed from the blood in one circuit. PAH (filtered and secreted), at the proper concentration, comes much closer to this. Approximately 10% of the blood passing through the normal kidney does not pass through functional nephrons (some goes to the renal capsule). Therefore, if one uses PAH to estimate renal plasma flow in a healthy subject, the estimate (effective renal plasma flow [ERPF]) will be lower than the true value but will be a good index of the "functional renal mass." In certain diseases of the kidney, the number of nonfunctional units may be increased, and the ERPF may be even more divergent from the true renal plasma flow.

446. (B)

$$CPAH = ERPF = (UPAH)/PPAH \times Vu$$
$$= 16 \text{ mg/mL of urine}/0.02 \text{ mg/mL of plasma}$$
$$\times 1 \text{ mL of urine/min}$$
$$= 800 \text{ mL of plasma/min}$$

447. (B)

448. (D) Any substance that is filtered by the glomeruli and has a renal clearance of 0 (as does glucose) must be actively reabsorbed. The ratio mentioned in this question equals clearance.

$$C = \text{rate of excretion/plasma concentration}$$
$$= \text{mg/min/mg/mL of plasma}$$
$$= \text{mL of plasma/min}$$

A renal clearance value lower than that of inulin could be due to either active or passive reabsorption, since inulin is filtered, but not reabsorbed or secreted. A renal clearance value higher than that of inulin is due to secretion, not reabsorption.

449. (E) The only substance listed that is not actively transported across the cells of the nephron and therefore does not have a transport maximum is urea.

450. (C) This ratio equals glomerular filtration rate. Ureteral obstruction increases the perivascular pressure for the glomerular capillaries and therefore decreases glomerular filtration. A decrease in glomerular capillary colloid osmotic pressure increases glomerular filtration.

451. (D) Inulin is filtered at the glomeruli, but not reabsorbed, secreted, or synthesized. Substances with a clearance greater than that for inulin must be moved into the nephron by a mechanism in addition to filtration. Filtration and secretion collectively can clear a greater quantity of blood than either alone. Agents that are both filtered and secreted but not reabsorbed (PAH and Diodrast, for example) have a clearance somewhat less than 5 times that for inulin (ie, their filtration fraction is about 0.2). A clearance 20 times that of inulin clearly indicates a mechanism in addition to filtration, secretion, and reabsorption. Reabsorption would result in a clearance less than that for inulin.

452. (E) Less urea is passively reabsorbed when less water is passively reabsorbed. In experiments in humans in which urine flow is modified by changing water intake, the following data are obtained: at a urine flow of 1 mL/min, 0.2 mM of urea were excreted per minute, and at a urine flow of 3 mL/min, 0.3 mM of urea were excreted per minute. Urea is filtered and passively reabsorbed but not secreted and therefore its clearance is less than that for inulin (filtered but not reabsorbed or secreted). Urea is formed in the liver.

453. **(B)**

454. **(E)** When increasing concentrations of a substance in the plasma cause a progressive decrease in the clearance of that substance, then the secretory Tm for that substance has been exceeded. Its large clearance at lower plasma concentrations probably means that it is also filtered.

455. **(C)** The progressive increase in clearance with increase in plasma concentration suggests active reabsorption. If substance h were actively secreted, there would be a progressive decrease in clearance with an increase in plasma concentration. A substance that is filtered and secreted, or filtered and synthesized, would have a larger clearance than what is shown for either substance h or g.

456. **(C)** The threshold is the lowest plasma concentration at which the ability of the transport system to move all of a particular substance across a barrier is exceeded. The transport maximum is reached somewhere beyond point j. The deviation between the threshold concentration and the transport maximum concentration is called the splay.

457. **(E)** In Tm-limited active transport (either secretion or reabsorption), many substances apparently compete with one another for sites on a carrier molecule. This is presumably the case with (1) PAH, phenol red, penicillin, and diodrast; (2) choline, histamine, and thiamine; (3) glucose, fructose, and phlorhizin; (4) the amino acids, arginine, lysine, and cystine; and (5) sulfate and thiosulfate. The changes listed in (A) and (B) would increase the excretion of penicillin if there were a carrier molecule for the reabsorption of penicillin or if penicillin were actively reabsorbed. If PAH acted to increase filtration, it would either produce no change or increase penicillin excretion.

458. **(B)** The drug probenecid blocks active reabsorption of uric acid in the kidney and therefore increases the loss of uric acid from the body. Uric acid is produced from the catabolism of nucleoproteins. If you wished to decrease the production of uric acid, a diet low in meats (particularly the nonmuscular meats, such as liver), meat extracts, and legumes would be far more effective than a diet low in carbohydrate. There is evidence that uric acid is both actively secreted and reabsorbed. To block active secretion would increase its concentration in the body.

459. **(A)**

$$\text{K excretion rate} = (\text{K concentration in urine}) \times (\text{urine production})$$

$$\text{K concentration in urine} = (100\ \text{mEq/L}) \times (1\ \text{L}/1000\ \text{mL}) = 0.1\ \text{mEq/mL}$$

$$\text{Urine production} = (1440\ \text{mL/day}) \times (1\ \text{day}/24\ \text{hr}) \times (1\ \text{hr}/60\ \text{min}) = 1\ \text{mL/min}$$

$$\text{K excretion rate} = (0.1\ \text{mEq/mL}) \times (1\ \text{mL/min}) = 0.1\ \text{mEq/min}$$

460. **(D)**

$$C_{ur} = (U_{ur})/P_{ur} \times V = (\text{urine concentration of urea})/(\text{plasma concentration of urea}) \times \text{urine flow}$$

$$= (2050\ \text{mg}/100\ \text{mL of urine})/(25\ \text{mg}/100\ \text{mL of plasma}) \times (1\ \text{mL of urine/min}) = 82\ \text{mL of plasma/min}$$

461. **(A)**

$$\text{Na reabsorbed} = (\text{Na filtered}) - (\text{Na excreted}) + (\text{Na secreted})$$

Na filtered: creatinine clearance is frequently used clinically to estimate the rate of glomerular filtration. The values obtained using exogenous creatinine are not an accurate nor reliable measure of filtration, but the values obtained in this case using endogenous creatinine yield data almost as good as one would get from an inulin clearance study:

$$C_{cr} = (U_{cr})/(P_{cr}) \times V$$
$$= (200 \text{ mg}/100 \text{ mL of urine})/(1 \text{ mg}/100 \text{ mL of plasma}) \times (1 \text{ mL of urine}/\text{min})$$
$$= 200 \text{ mL of plasma}/\text{min}$$
$$\text{Na filtered} = (\text{Na concentration in plasma}) \times (\text{glomerular filtration rate})$$
$$= (0.140 \text{ mEq}/\text{mL}) \times (200 \text{ mL}/\text{min})$$
$$= 28 \text{ mEq}/\text{min}$$
$$\text{Na excreted} = (\text{Na concentration in urine}) \times (\text{urine production})$$
$$= (0.120 \text{ mEq}/\text{mL}) \times (1 \text{ mL}/\text{min})$$
$$= 0.120 \text{ mEq}/\text{min}$$
$$\text{Na secreted} = 0$$
$$\text{Na reabsorption} = (28 \text{ mEq}/\text{min}) - (0.12 \text{ mEq}/\text{min}) = 27.88 \text{ mEq}/\text{min}$$

462. (D)

$$\text{Fraction of urea excreted} = (\text{urea excreted})/(\text{urea filtered})$$
$$\text{Urea excreted} = (\text{urea concentration in urine}) \times (\text{urine production})$$
$$= (20.5 \text{ mg}/\text{mL}) \times (1 \text{ mL}/\text{min}) = 20.5 \text{ mg}/\text{min}$$
$$\text{Urea filtered} = (\text{urea concentration in plasma}) \times (\text{plasma filtered})$$
$$= (0.25 \text{ mg}/\text{mL}) \times (200 \text{ mL}/\text{min})$$
$$= 50 \text{ mg}/\text{min}$$
$$\text{Fraction of urea excreted} = (20.5 \text{ mg}/\text{min})/(50 \text{ mg}/\text{min}) = 0.41$$

463. (B) This apparently does not occur. The Na^+ current is into the descending limb. In countercurrent multiplication, (1) the ascending limb by virtue of its impermeability and active transport system creates (2) a hypernatremic environment in the peritubular fluid, which causes (3) a hypernatremia in the lumen of the permeable descending limb, which causes (4) hypernatremia in the ascending limb, which permits (5) an even greater transfer of Na^+ from the ascending limb into the peritubular fluid.

464. (E) The loop of Henle and the medullary collecting duct are permeable to water. Antidiuretic hormone increases the permeability of the collecting duct to water. Both are responsible for the hypertonicity of the medullary peritubular compartment. Both contain a hypertonic solution in the presence of ADH. The ADH increases the osmotic pressure of the fluid in the medullary collecting duct by increasing its permeability to water (ie, by increasing water reabsorption).

465. (E) The juxtamedullary nephrons constitute less than 15% of the nephrons. The cortical nephrons constitute the rest. The cortical nephrons have a short loop of Henle that extends into the outer zone of the medulla. The juxtamedullary nephrons, like the collecting ducts, extend well into the inner zone of the medulla. Some have loops of Henle that extend to the papillary tip of the medulla. Both types of nephrons have their glomerulus in the cortex and their loop of Henle in the medulla.

466. (D) Normally, the active transport of NaCl from the lumen of the proximal tubule is responsible for a diffusion of water from the lumen that causes a fourfold increase in inulin concentration in the luminal fluid. The net result is that the luminal fluid has an osmotic pressure and Na^+ concentration similar to that in the glomerular filtrate. In other words, NaCl reabsorption in the proximal tubule is a mechanism for the concentration of inulin and other poorly reabsorbed substances. Normally, the inulin concentration would be 400% higher than that in plasma, and the Na^+ concentration would be about 150 mM/L. Decreases in filtration will not cause large changes in inulin concentration. Inulin is neither secreted nor reabsorbed. The active transport of urea has not been demonstrated in humans.

467. (A) Approximately 70% of the water in the glomerular filtrate is reabsorbed in the proximal tubule. This is due to (1) the permeability of the tubule to water and (2) the osmotic gradient between the tubular and peritubular fluid. Abnormally high concentrations in the glomerular filtrate of substances that are not readily reabsorbed (glucose in the case of hyperglycemia) prevent the maintenance of this gradient and therefore decrease the reabsorption of water in the proximal tubule. In water diuresis, there is also a reduced reabsorption of water, but this occurs at the distal tubule and collecting duct, not at the proximal tubule. Potassium excretion will be little affected by diuresis. Urea concentration in the urine is decreased during diuresis. If the diuresis causes an increase in the tonicity of the blood, there will be an increased release of ADH if the hypothalamus and pituitary are functioning normally. A decrease in ADH release is one mechanism that produces diuresis in response to a water load.

468. (B) The major fraction of urinary ammonia is derived from the amide nitrogen of glutamine.

469. (E) The transport maximum (Tm) is the greatest rate that a transport system can move a solute. It occurs at a concentration of solute that saturates the transport system. The carriers for Na^+ elsewhere in the nephron do not saturate at the Na^+ concentrations to which they are exposed.

470. (B) The high level of Na^+ reabsorption and the high permeability to water combine in the proximal tubule to move large volumes of water toward the blood.

471. (D) The ascending limb is also poorly permeable to urea, and an area of active Na^+ reabsorption. The Na^+ reabsorption contributes to the hypertonicity of the peritubular space in the renal medulla. The ADH (antidiuretic hormone = vasopressin) increases the permeability of the collecting duct to water. In the absence of ADH, the collecting duct is impermeable to water.

472. (C) It is surrounded by a hypertonic solution and is permeable to water.

473. (F) Vasopressin (= ADH = antidiuretic hormone) increases the permeability of the collecting duct to water and urea. There results a reabsorption of water and a more concentrated urine.

CHAPTER 20

Acid–Base Balance

Questions

DIRECTIONS (Questions 474 through 485): Each set of items in this section consists of a list of lettered options followed by several numbered words or phrases. For each numbered word or phrase, select the ONE lettered option that is most closely associated with it. Each lettered option may be selected once, more than once, or not at all.

Questions 474 through 477

(A) normal pH and high bicarbonate concentration in the blood plasma
(B) near normal pH and a $[HCO_3^-]/[H_2CO_3]$ ratio much lower than 20:1
(C) abnormally low blood pH due to excess fixed acid
(D) abnormally low pH and a high bicarbonate concentration in the plasma
(E) abnormally high blood pH

474. A definition of uncompensated metabolic acidosis.

475. A definition of compensated metabolic acidosis.

476. A definition of uncompensated respiratory acidosis.

477. A definition of compensated respiratory acidosis.

Questions 478 through 481

Each of the following questions has either two or three correct matches.

(A) high plasma pH
(B) low plasma pH
(C) near normal plasma pH
(D) normal plasma pH
(E) high plasma $[HCO_3^-]$
(F) low plasma $[HCO_3^-]$
(G) a urine pH between 6.5 and 4.5

478. Sign of metabolic alkalosis.

479. Sign of compensated metabolic alkalosis.

480. Sign of respiratory alkalosis.

481. Sign of compensated respiratory alkalosis.

Questions 482 through 485

Some of the following questions have more than one correct match.

(A) asphyxia
(B) diabetes mellitus
(C) hyperventilation
(D) ingestion of $NaHCO_3$
(E) regurgitation of gastric contents
(F) strenuous running
(G) ingestion of vegetables and fruit
(H) ingestion of proteins
(I) chronic renal disease

482. Cause of metabolic acidosis.

483. Cause of metabolic alkalosis.

484. Cause of respiratory acidosis.

485. Cause of respiratory alkalosis.

DIRECTIONS (Questions 486 through 489): Each of the numbered items or incomplete statements in this section is followed by answers or by completions of the statement. Select the ONE lettered answer or completion that is BEST in each case.

486. The addition of a fixed acid such as H_2SO_4 to the blood during protein catabolism may produce little change in the pH of the blood because of the blood buffer systems. Which buffer system in the healthy subject will usually combine with the most H^+?

(A) $NaHCO_3$-H_2CO_3
(B) Na_2HPO_4-NaH_2PO_4
(C) hemoglobin
(D) plasma proteins
(E) Na-K

487. Renal bicarbonate reabsorption

(A) reaches a plasma threshold at about the resting plasma bicarbonate concentration
(B) has a transport maximum that is elevated by increases in P_{CO_2} and aldosterone
(C) A and B
(D) has a stable Tm
(E) is not active

488. A 65-year-old woman complaining of dyspnea enters the hospital. The following data are obtained from her and from a normal healthy subject:

	PATIENT	NORMAL
Arterial P_{O_2} (mm Hg):	75	95
Arterial P_{CO_2} (mm Hg):	55	40
Arterial hemoglobin (g/100 mL):	18	15
Arterial pH:	7.35	7.40

The patient is suffering from

(A) respiratory alkalosis
(B) respiratory acidosis
(C) metabolic alkalosis
(D) metabolic acidosis
(E) anemia

489. The rate of secretion of H^+ by the nephron is

(A) greater in the proximal tubule than in the distal tubule
(B) increased in response to aldosterone
(C) A and B
(D) increased when plasma K^+ increases
(E) inversely related to the rate of reabsorption of bicarbonate

Answers and Explanations

474. (C) Fixed acids are nonvolatile H^+ donors, such as lactic acid and H_2SO_4, and the acid salts, NH_4Cl and NaH_2PO_4.

475. (B) In metabolic acidosis, the respiratory system responds to the increased $[H^+]$ by increasing ventilation. The net result is a decreased P_{CO_2} and a marked decrease in $[HCO_3^-]$, because of the decreased P_{CO_2} and because H^+ drives the following equilibrium to the right:

$$H^+ + HCO_3^- \leftrightarrow H_2CO_3$$

As a result the $[HCO_3^-]/[H_2CO_3]$ ratio changes from 20 (at a pH of 7.4) to a much lower number (1.1 at a pH of 6.06).

476. (D) In this condition the retention of the acid-forming gas, CO_2, causes the decreased pH and the increased bicarbonate.

477. (A) Inadequate ventilation causes an increased plasma P_{CO_2}, $[H^+]$, and $[HCO_3^-]$. The kidney is able to return the $[H^+]$ to normal by increasing the excretion of H^+ and a number of H^- containing salts. These salts (NaH_2PO_4, $NaHSO_4$, NH_4Cl) in a sense "smuggle" the H^+ out of the body without as markedly affecting urine pH as H^+ plus Cl^- do. The kidney is generally much better, although slower, in compensating for a respiratory acidosis or alkalosis than the respiratory system is for compensating for a metabolic pH change. This is because the respiratory system uses only one mechanism (ventilation) to control four different parameters (P_{O_2}, P_{CO_2}, $[H^+]$, $[HCO_3^-]$). Therefore, in bringing one parameter (H^+) into balance, it upsets the balance of one or more other parameters (CO_2 and HCO_3^- for example). When we decrease $[H^+]$ by increasing ventilation, we also decrease P_{CO_2}. Since a decreased P_{CO_2} causes a decreased ventilation, the decreased P_{CO_2} prevents the compensation for the high $[H^+]$ from being complete:

$$\text{Increased } [H^+] \rightarrow \text{Increased ventilation} \rightarrow \text{Decreased } [H^+] \rightarrow \text{Decreased ventilation Decreased } P_{CO_2}$$

Under healthy, resting conditions, the plasma bicarbonate is about 24 mM/L, but during respiratory acidosis or metabolic alkalosis it may rise to 40 mM/L.

478. (A, E, G) A high alkaline ingestion ($NaHCO_3$ or vegetables) causes an alkaline urine. Renal failure can cause an acid urine. Alkalosis drives the following reaction in the plasma to the right:

$$OH^- + H_2CO_3 \leftrightarrow HOH + HCO_3^-$$

479. (C, E) The $[HCO_3^-]$ is increased further by the compensation (ie, decreased ventilation).

480. (A, F, G) The kidney is not able to produce a urinary pH below 4.5 or above 8.0. Hyperventilation causes respiratory alkalosis by eliminating an excess of the acid-forming gas, CO_2, and by so doing decreases plasma $[HCO_3^-]$ as well as $[H^+]$.

481. **(D, F)** Renal compensation (increased excretion of bicarbonate, decreased excretion of H^+ and acid salts) exaggerates the decreased bicarbonate concentration, but is usually able to bring the elevated pH back to normal. The kidney, unlike the lung, is usually able to change the plasma pH without upsetting the other parameters it controls (water balance, electrolyte balance, and the elimination of urea, for example). The kidney responds to respiratory alkalosis by excreting more $NaHCO_3$ or H_2CO_3. Which molecule is eliminated in greatest quantity will be determined by the body's need to conserve Na or K. In respiratory acidosis the kidney eliminates NH_4Cl, NaH_2PO_4, or KH_2PO_4. Which molecule is eliminated is determined by the body's need to conserve Na, K, or both. The NH_4^+, unlike Na^+ and K^+, is only important as a part of the kidney's mechanism for maintaining acid–base balance.

483. **(B, F, H, I)** Metabolic acidosis may be caused by an increased production of acid (ketone bodies, lactate, etc.) or an increased intake of OH^-, removing salt or acid-forming nutrients, or a decreased excretion of fixed acid. The ingestion of NH_4Cl drives the following reaction to the right:

$$OH^- + NH_4Cl \leftrightarrow NH_4OH + Cl^-$$

In the healthy subject on an average U.S. diet, the kidneys excrete 40 to 80 mEq of nonvolatile acid per day and the lungs eliminate about 13,000 mEq of volatile acid (CO_2) per day. In chronic renal disease the kidney has a reduced ability to produce NH_3 and therefore a reduced ability (1) to eliminate H^+ (ie, the urine contains less NH_4Cl) and (2) to prevent a highly acid urine (ie, the urine is less well buffered and therefore may approach a pH of 4.5).

483. **(D, E, G)** Metabolic alkalosis may be caused by the ingestion of an alkaline salt or nutrient, or the loss of an acid secretion (gastric juice).

484. **(A)** Asphyxia equals hypoxia + hypercapnea. The hypercapnea causes the acidosis.

485. **(C)** Hyperventilation decreases plasma P_{CO_2}.

486. **(C)** Hemoglobin, under these circumstances, will combine with the most H^+. Other systems that buffer changes in H^+ are plasma proteins, the $NaHCO_3$-H_2CO_3 pair (the principal buffer system in the entire extracellular fluid), and the Na_2HPO_4-NaH_2PO_4 pair.

487. **(C)**

488. **(B)** When the pH is low (= 7.35) and the P_{CO_2}, HCO_3^-, or H_2CO_3 are high (P_{CO_2} = 55 mm Hg in the arteries) the condition is respiratory acidosis.

489. **(C)** An active transport system apparently moves H^+ into the tubular lumen. Aldosterone facilitates the conservation of Na^+ and the excretion of H^+ and K^+. Hydrogen ion secretion is inversely related to plasma K^+ concentration. It is directly related to bicarbonate reabsorption.

CHAPTER 21

Control of the Kidney

Questions

DIRECTIONS (Questions 490 through 503): Each of the numbered items or incomplete statements in this section is followed by answers or by completions of the statement. Select the ONE lettered answer or completion that is BEST in each case.

490. A decreased pressure in the carotid sinus causes

(A) an increased renal adrenergic sympathetic tone
(B) renal vasoconstriction
(C) A and B
(D) an increased secretion of renin by the juxtaglomerular cells
(E) A, B, and D

491. In a series of experiments, one group of animals has both kidneys removed and a second group has a sham operation. Each group is then bled until the arterial pressure decreases to 10 mm Hg. The animals are then permitted 40 minutes to recover. At the end of this period, it is found that the sham nephrectomized animals have an arterial pressure 30 mm Hg higher than the nephrectomized animals. Which of the following hypotheses do these observations support? In response to hemorrhage, the kidney

(A) (by conserving water) can increase the arterial pressure
(B) (by producing an agent that either directly or indirectly causes an increased systemic resistance) increases the arterial pressure
(C) does both A and B
(D) releases a vasodilator agent
(E) does both A and D

492. What is the agent released by the kidney in Question 491 that tends to counter the hypotensive action of hemorrhage?

(A) renin
(B) angiotensin
(C) aldosterone
(D) epinephrine
(E) norepinephrine

493. What actions does angiotensin II have that are not related to its ability to produce vasoconstriction? Angiotensin II

(A) increases extracellular fluid volume
(B) increases K^+ and H^+ excretion
(C) A and B
(D) decreases the reabsorption of Na^+ by the kidney
(E) A, B, and D

494. Which of the following statements is INCORRECT? The antidiuretic hormone (ADH)

(A) is released in response to a hypertonic solution infused into the arteries serving the anterior hypothalamus
(B) decreases blood volume
(C) is released during water deprivation
(D) is released in the well-hydrated subject
(E) is released in response to stimulation of the supraoptic nuclei of the hypothalamus

495. Which of the following changes does NOT produce polyuria?

(A) diabetes mellitus
(B) glucosuria
(C) tissue refractoriness to insulin
(D) diabetes insipidus
(E) aldosterone injection

496. Which of the following statements is INCORRECT? There will be an increased release of aldosterone from the adrenal cortex in response to

(A) an increased arterial pressure
(B) hyponatremia
(C) adrenocorticotropic hormone (ACTH) from the anterior pituitary gland
(D) hyperkalemia
(E) epinephrine and norepinephrine

497. Which of the following statements is INCORRECT? Parathyroid hormone

(A) can cause an increased muscle tone and hyperreflexia
(B) decreases the renal absorption of phosphate
(C) increases the renal absorption of Ca^{++}
(D) can cause a demineralization of bone
(E) can cause an increase in urinary Ca^{++}

498. A patient is noted to have a markedly reduced plasma colloid osmotic pressure and Na^{+} excretion. After the intravenous infusion of a concentrated albumin solution, the patient has a rapid weight loss. The signs, prior to treatment, are consistent with a diagnosis of

(A) edema
(B) an elevated aldosterone secretion
(C) A and B
(D) a low secretion of antidiuretic hormone
(E) A and D

499. In Question 498, the infusion of a concentrated albumin solution

(A) decreased renin production
(B) increased mean arterial pressure
(C) A and B
(D) increased production of angiotensin II
(E) A, B, and D

500. If there is a decrease in extracellular fluid volume, the body may respond by increasing the secretion of

(A) norepinephrine
(B) aldosterone
(C) norepinephrine and aldosterone
(D) ADH
(E) norepinephrine, aldosterone, and ADH

501. Which of the following statements is INCORRECT?

(A) the distention produced by urine flowing into the minor renal calyces triggers a calyceal contraction
(B) the renal pelvis undergoes rhythmic dilation and contraction
(C) after closure of the pelviureteric junction, a peristaltic wave in the ureter carries urine toward the bladder
(D) as the bladder fills, there is a progressive increase in intravesicular pressure similar to what one finds when a balloon is filled with air
(E) the stimulation of autonomic neurons originating from S2, S3, and S4 will initiate the contraction of the urinary bladder

502. The residual volume for an organ is the volume of fluid remaining in its lumen after a maximal contraction. The approximate residual volume in the adult for the left ventricle is 25 mL and for the lungs is 1400 mL. The residual volume for the urinary bladder is

(A) between 10 and 20 mL in the healthy subject
(B) lowered after a transection of the cord at L3
(C) lowered in the late stages of tabes dorsalis where there is sclerosis of the posterior columns and spinal roots
(D) all of the above
(E) none of the above

503. Which of the following statements is INCORRECT?

(A) contraction of the detrusor muscle widens the posterior urethra
(B) urine remaining in the urethra of the male at the end of micturition is expelled by the contraction of the bulbocavernosus muscle
(C) voiding can be voluntarily terminated by the contraction of the external urethra sphincter
(D) urine flow can be accelerated by the Valsalva maneuver
(E) the decentralized bladder is unable to contract

Answers and Explanations

490. **(E)** When the pressure in the arterial or cardiopulmonary stretch receptors is reduced, there is an increased renal, adrenergic, sympathetic tone that causes an increased resistance to renal blood flow and an increased secretion of renin. In the decentralized or transplanted kidney, most renal functions seem to remain normal, since the autoregulatory mechanisms that control renal vasoconstriction and release of renin remain intact.

491. **(B)** This observation is consistent with the data presented. The kidney conserves water by decreasing urine volume. In the nephrectomized animals there would be no urine production, so, on this basis, the nephrectomized animals would not have a lower pressure after hemorrhage. A vasodilator would decrease arterial pressure.

492. **(A)** It is renin, and it is produced by juxtaglomerular cells surrounding the renal afferent arterioles.

493. **(C)** Angiotensin II facilitates the release of aldosterone from the adrenal cortex. Aldosterone, by causing the kidney to reabsorb more osmotically active material (Na^+ and Cl^-) than it excretes, increases extracellular fluid volume (see Fig. 21–1).

494. **(B)** ADH secretion rises during water deprivation and falls during water loading. It is released in varying quantities under most circumstances (see Fig. 21–2).

495. **(E)** The increased reabsorption of Na^+ induced by aldosterone causes an increased reabsorption of H_2O (osmotic effect). One cause of diabetes mellitus is tissue refractoriness to insulin:

Reduced insulin action → Increased plasma glucose → Glucosuria → Polyuria (osmotic effect of glucose)

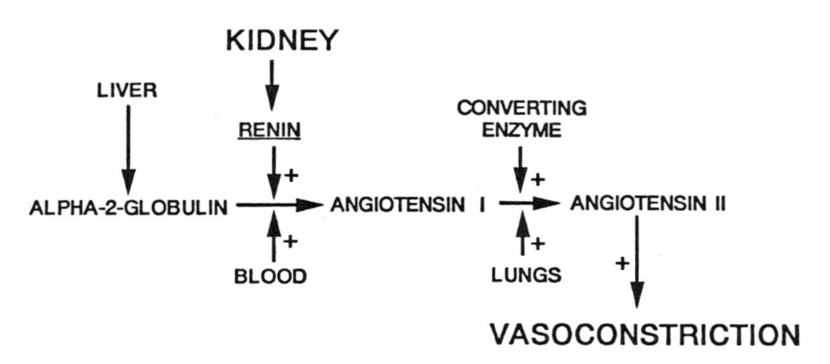

Figure 21–1. Actions of angiotensin II in renal function.

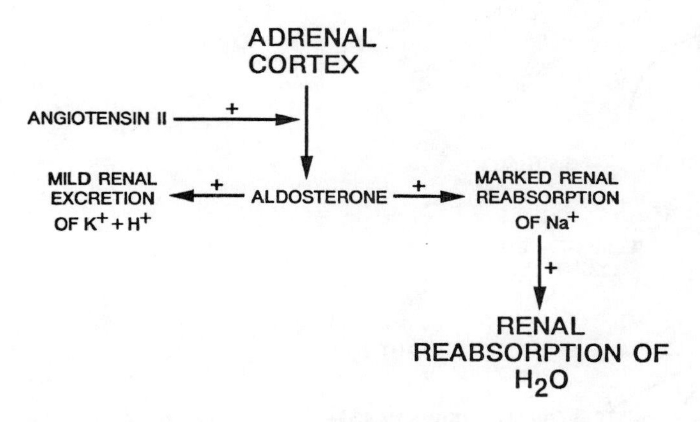

Figure 21–2. Actions of antidiuretic hormone (ADH) on kidney function.

In diabetes insipidus, the polyuria is due to inadequate quantities of ADH and, hence, reduced reabsorption of water in the kidney.

496. **(A)** Increases in arterial pressure decrease the secretion of aldosterone (see Fig. 21–3).

497. **(A)** Parathormone (PTH) produces an increased plasma Ca^{++} (see Fig. 21–4 on page 188). It is a decreased plasma Ca^{++} that causes tetany (increased muscle tone and hyperreflexia).

498. **(C)** A decreased ADH secretion would cause dehydration, not edema, and would not reduce the Na^+ excretion (see Fig. 21–5 on page 188).

499. **(C)** See Figure 21–6 on page 188.

500. **(E)** Each of these agents increases arterial pressure by either increasing cardiac output and peripheral resistance or by increasing blood volume.

501. **(D)** The urinary bladder, unlike a balloon, accommodates for increases in volume by adjusting its tone. The injection of 100 mL of liquid into the bladder causes an initial increase in pressure of about 3 cm H_2O, but this is followed by a decrease in pressure (accommodation). It is not until the volume of the adult bladder exceeds 250 to 450 mL that intravesicular pressure exceeds 5 cm H_2O. During micturition, the pressure in the bladder exceeds 30 cm H_2O. Pressure in the upper ureter averages 15 cm H_2O and in the pelvic ureter averages about 30 cm H_2O during ureteral systole.

502. **(E)** Under normal conditions, micturition leaves a residual volume of 0.09 to 2.34 mL. Since urine is a medium that supports the

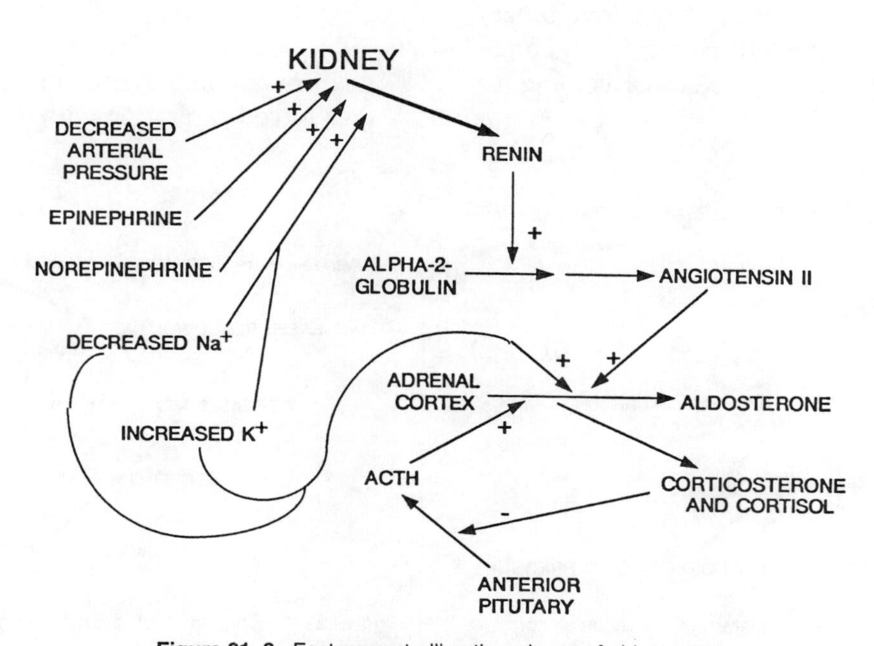

Figure 21–3. Factors controlling the release of aldosterone.

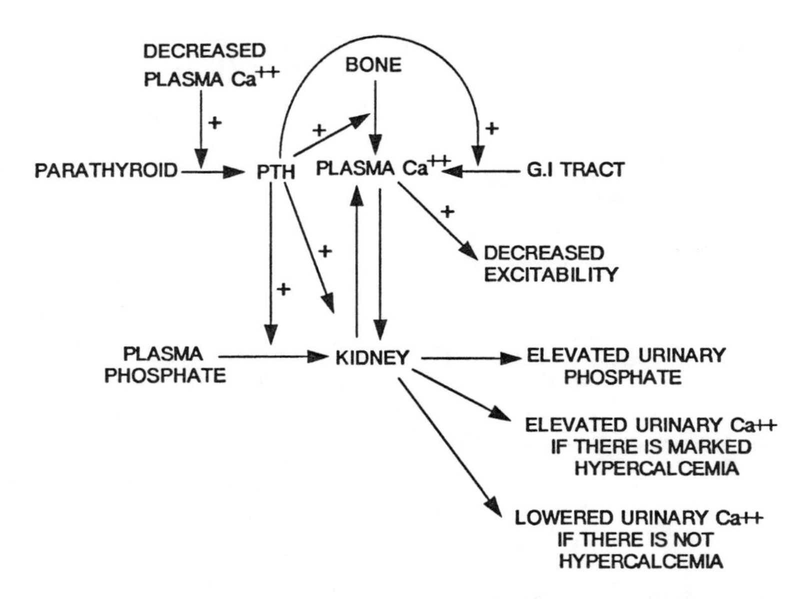

Figure 21–4. Role of parathyroid hormone in mobilization of Ca and phosphate.

growth of microorganisms, the higher urine residual volumes seen after certain central nervous system disorders may lead to infection of the urinary system which, at its worst, can be fatal.

503. (E) The decentralized bladder is initially distended and flaccid but eventually attains a small volume, contracts frequently in response to distention, but has contractions of short duration. This results in the frequent involuntary expulsion of small volumes of urine and a potentially dangerous large residual volume for the bladder.

A number of physiologists persist in referring to the neck of the bladder as the internal urethral sphincter. There is little anatomic or physiologic data to support this thesis. Apparently, the contraction of the detrusor muscle reduces the resistance to outflow by both shortening and widening the posterior urethra (= prostatic urethra in the male). In short, there is little to indicate that in micturition there is an "inhibition of the internal urethral sphincter."

In the Valsalva maneuver, an increased intra-abdominal pressure is produced during a forced expiration with the glottis closed.

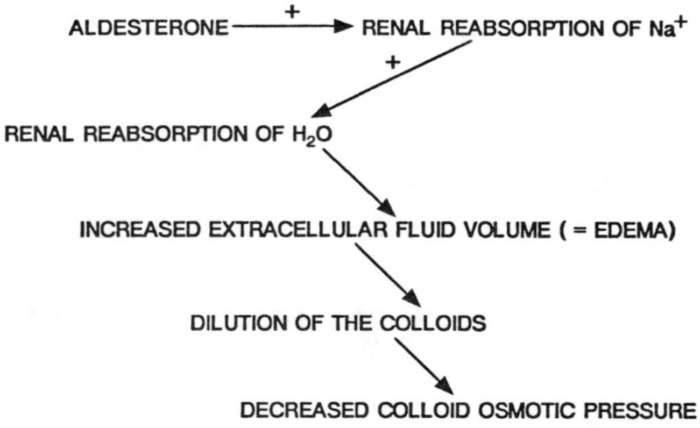

ALDESTERONE —+→ RENAL REABSORPTION OF Na^+

+

RENAL REABSORPTION OF H_2O

INCREASED EXTRACELLULAR FLUID VOLUME (= EDEMA)

DILUTION OF THE COLLOIDS

DECREASED COLLOID OSMOTIC PRESSURE

Figure 21–5. Effects of excess aldosterone on vascular components.

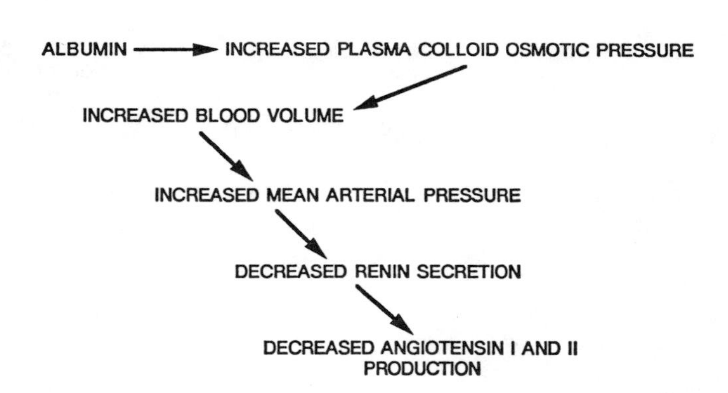

Figure 21–6. Sequence of events when plasma protein is increased.

PART VII

Digestion and Energy Balance

CHAPTER 22

Mouth, Esophagus, and Stomach

Questions

DIRECTIONS (Questions 504 through 521): Each of the numbered items or incomplete statements in this section is followed by answers or by completions of the statement. Select the ONE lettered answer or completion that is BEST in each case.

504. Which of the following statements is MOST correct?

(A) the secretion of saliva is primarily under hormonal control
(B) the secretion of saliva is increased by parasympathetic stimulation
(C) the ratio of salivary/plasma iodine concentration lies between 0.8 and 2.0
(D) the concentration of Na^+ in the saliva is increased by aldosterone
(E) the increased concentration of ketone bodies in the plasma in diabetes mellitus does not affect their concentration in the saliva

505. Which of the following statements is INCORRECT? Saliva

(A) is essential for the complete digestion of starch
(B) prevents dental caries
(C) prevents decalcification of the teeth
(D) is a well-buffered solution that tends to maintain a pH of about 7.0 in the mouth
(E) produced by the serous gland cells of the mouth contains a higher concentration of salivary amylase and a lower concentration of mucin than the saliva from the mucous gland cells of the mouth

506. Which of the following statements is INCORRECT?

(A) removal of the epiglottis frequently causes aspiration pneumonitis
(B) elevation of the larynx is an important part of volitional swallowing
(C) liquids swallowed by an erect individual may reach the esophagogastric junction while it is still closed
(D) the esophagogastric junction usually prevents regurgitation of the gastric contents into the esophagus, and in this way tends to prevent heartburn
(E) food lodged in the esophagus initiates an involuntary reflex that carries the food to the stomach, but does not include elevation of the larynx

507. Which of the following statements is MOST correct?

(A) some of the fat in ingested food may remain in the stomach for as long as 5 hours
(B) water in the stomach is more slowly absorbed than water in the intestine
(C) pancreatic juice has a higher concentration of bicarbonate than gastric juice
(D) the pyloric sphincter remains open most of the time
(E) all of the above are correct

508. Which of the following statements is INCORRECT? The stomach

(A) decreases its motility in response to H^+ in the duodenum
(B) decreases its motility in response to fatty acids in the duodenum
(C) decreases its motility in response to the hormone enterogastrone
(D) is essential for vomiting in the adult
(E) decreases its motility in response to an enterogastric reflex

509. Gastric motility is accelerated by

(A) a hyperosmolality of the duodenal contents
(B) the products of protein digestion in the duodenum
(C) distention of the duodenum
(D) A, B, and C
(E) a decreased secretion of cholecystokinin

510. A pH below 3.5, fats, and hypertonicity in the duodenum decrease gastric motility by

(A) a reflex stimulation of sympathetic, efferent neurons in the thoracolumbar cord
(B) a stimulation of an intramural, intrinsic plexus in the duodenum, which can depress gastric motility after decentralization of the GI tract
(C) A and B
(D) decreasing the release of secretin
(E) A, B, and D

511. Which of the following statements is INCORRECT? The stomach

(A) depends on carbonic anhydrase for the production of hydrochloric acid
(B) decreases its mechanical activity in response to epinephrine
(C) absorbs over 25% of the products of protein catabolism produced in the GI tract
(D) produces and releases gastrin, somatostatin, glucagon, and intrinsic factor
(E) has less contractile activity one-half hour after a meal of solid food than when it is empty

512. Which of the following statements is INCORRECT? Hunger contractions in the stomach

(A) are not essential for the sensation of hunger
(B) may be caused by a reduction of blood glucose levels
(C) may be caused by a reduction of blood glucose levels in the patient who has had a bilateral, supradiaphragmatic vagotomy
(D) are inhibited by hyperglycemia
(E) occur in diabetes mellitus

513. Which of the following statements is INCORRECT? The stomach

(A) responds to an increase in its contents from 600 to 1600 mL with less than a 5 mm Hg increase in pressure
(B) empties more rapidly in response to a liquid meal than a solid meal
(C) during emptying, produces a more forceful contraction in its pyloric portion than in its body
(D) contains a pacemaker area near its cardiac portion
(E) secretes the following enzymes: pepsin, trypsin, lipase, amylase

514. Which of the following statements is INCORRECT? Vomiting in the adult

(A) begins with salivation and a sense of nausea
(B) is controlled by a center in the reticular formation of the medulla
(C) in response to apomorphine, a number of other emetic drugs, radiation, and uremia can be abolished by lesions in the area postrema
(D) in response to irritation of the mucosa in the upper GI tract can be abolished by lesions in the area postrema
(E) is caused by visceral afferent neurons in sympathetic nerves and the vagi

515. In the healthy newborn infant, as opposed to the adult,

(A) skeletal muscle contraction is more important in causing vomiting
(B) the esophagogastric junction is less effective in preventing regurgitation
(C) the gastric juice is less acidic
(D) heartburn (painful irritation of the esophagus) is less frequent
(E) regurgitated gastric contents bubble out the mouth more often

516. Which of the following statements is INCORRECT? In the

(A) stomach, pepsinogen is activated to pepsin by HCl
(B) stomach, pepsin is formed and, in the duodenum, is inactivated by the higher pH
(C) stomach, in the pyloric gland area, there is a complete replacement of cells every 2 days
(D) stomach, mucus forms a lining greater than 1 mm thick over the membranes and keeps their pH near 7.0
(E) newborn infant, the stomach is less permeable than the stomach of the adult

517. Which of the following statements is MOST correct? After the stomach is removed, the patient who receives no further treatment will usually develop

(A) symptoms due to a decreased absorption of iron
(B) symptoms due to a decreased absorption of vitamin B_{12}
(C) an anemia due to the decreased absorption of iron and vitamin B_{12}
(D) a carbohydrate loss in the feces that is in excess of 25% of the normally digestible carbohydrate intake (ie, excluding cellulose and bran)
(E) all of the above

518. Which of the following statements is MOST correct? After the stomach is removed, the patient who receives no further treatment will usually develop

(A) exaggerated fluctuations in plasma glucose concentration after a meal
(B) exaggerated fluctuations in blood volume
(C) marked increases in the lipid content of the feces
(D) A, B, and C
(E) increased secretion of pancreatic juice

519. Which of the following statements is MOST correct? Supradiaphragmatic, bilateral vagal section prevents gastric secretion in response to

(A) feelings of hostility
(B) distention of the stomach
(C) feelings of hostility and distention of the stomach
(D) the products of protein catabolism in the duodenum
(E) all of the above

520. In 1964, Gregory and Tracey purified gastrin. It is best characterized as a linear peptide that

(A) is released by the stomach in response to the stimulation of preganglionic parasympathetic neurons to the stomach
(B) is released by the stomach in response to the distention of the stomach
(C) is released in response to the products of protein digestion in the duodenum
(D) increases gastric secretion
(E) all of the above

521. Acid secretion in the stomach is inhibited by

(A) an H1 histamine blocker (benadryl)
(B) an H2 histamine blocker (cimetidine)
(C) atropine
(D) A and C
(E) B and C

Answers and Explanations

504. (B) Although the content of the saliva may vary with changes in hormonal concentrations, the major way the body initiates the secretion of saliva is through the stimulation of parasympathetic neurons to the salivary glands. The ratio of salivary/plasma iodine concentration sometimes reaches 60. The concentration of Na^+ in the saliva is decreased by aldosterone. Increased plasma concentrations of ketone bodies, urea, and mercury can also cause increases of these substances in the saliva.

505. (A) Pancreatic amylase can digest starch in the absence of salivary amylase. Patients with a deficient secretion of saliva have a higher incidence of caries than normal subjects. The saliva, by keeping the pH of the mouth from becoming acid, prevents the decalcification of the teeth.

506. (A) It is the elevation of the larynx that prevents the aspiration of solids and liquids. Voluntary swallowing becomes very difficult if the larynx is immobilized or if one attempts to swallow while the mouth remains open. Each of these events is part of a series that eventually leads to the delivery of food or water to the stomach. Normally, the esophagogastric junction will open before the peristaltic wave reaches it, but because liquids move downhill more rapidly than peristalsis, they sometimes reach the junction while it is still closed. The phenomenon in option (E) is sometimes called secondary peristalsis.

507. (E) Fat slows gastric emptying. Water placed in the intestine is absorbed in about 10 minutes. Water in the stomach is absorbed in about 60 minutes. Gastric juice has a lower bicarbonate concentration than any of the alkaline secretions of the alimentary tract (ie, saliva, pancreatic juice, bile, and the succus entericus).

508. (D) In the adult, unlike the infant, vomiting is more frequent in the absence of a stomach than in its presence. In the adult, vomiting is caused primarily by a build-up of pressure in the abdomen in response to a closure of the glottis and a contraction of the diaphragm and other skeletal muscles. H^+ and the products of protein digestion in the duodenum decrease gastric motility by means of the enterogastric reflex. Fatty acids, triglycerides, and phospholipids act on the mucosa of the duodenum to cause the release of enterogastrone, which is carried in the blood to the stomach where it inhibits gastric motility and thus delays the emptying of the stomach. An enterogastric reflex inhibits vagal parasympathetic tone to the stomach.

509. (E) Duodenal fats cause the release of cholecystokinin. Options (A), (B), and (C) decelerate gastric emptying.

510. (C) Unlike most neurons, those in the intrinsic plexus can control and integrate contractile function after all the neurons coming from and going to the central nervous system have been cut or blocked. Secretin is released in response to a low duodenal pH and decreases gastric motility, while also increasing the secretion of bile, pancreatic juice, and duodenal juice into the duodenum.

511. (C) Less than 15% of the protein that enters the stomach is catabolized to amino acids. The catabolism of proteins that occurs in the stomach is only the first step in protein catabolism. Removal of the stomach or its failure to produce HCl does not affect the nitrogen content of the feces. After 3 hours of fasting, the contractions are more frequent, more forceful, and of longer duration than one-half hour after a meal.

512. (C) It is hypothesized that there are a number of cells within the satiety center of the hypothalamus that decrease the activity of the center as the quantity of available glucose diminishes. These cells collectively are called glucostats. In hypoglycemia, they initiate the sensation of hunger and, through vagal parasympathetic neurons, hunger contractions. In other words, hunger contractions are not thought of as the cause of the hunger sensation but are merely associated with hunger. On the other hand, they may be so strong as to cause a painful sensation (hunger pangs) in addition to the sensation of hunger.

In diabetes mellitus there is hyperglycemia, but the blood sugar is not readily available for cell metabolism. Therefore, the glucostat responds to this condition as it would to hypoglycemia. There are many situations such as this in the body. We have noted, for example, that in carbon monoxide poisoning the important factor in oxygen transport is not the concentration of hemoglobin, but rather the concentration of *available* hemoglobin. This perspective is also important in considering the action of hormones bound to plasma proteins (eg, thyroxine).

513. (E) The stomach does not secrete trypsin or amylase. The digestion of starch that occurs in the stomach is due to salivary amylase. The stomach, like the urinary bladder, shows receptive relaxation. An increase in gastric volume might very well produce no change in intragastric pressure. The pyloric part of the stomach is more muscular than the more cephalad portions. The contractions in the body serve to mix the gastric contents with enzymes and acid. The contractions in the pyloric antrum serve to move liquid chyme into the intestine and solid material into the body.

514. (D) The area postrema has been hypothesized to contain a chemoreceptor trigger zone that is sensitive to certain agents, such as apomorphine, and that sends stimuli to the vomiting center. Sensory impulses from the GI tract do not pass through the trigger zone.

515. (A) Vomiting or regurgitation in the infant is not prevented by skeletal muscle blockade, whereas in the adult it is. In adults, skeletal muscle contraction may be so marked during vomiting that there is pain. In infants, vomiting is almost totally a gastroesophageal phenomenon and is seldom if ever painful. The higher pH of the infant's gastric contents protects him from heartburn during regurgitation, since heartburn occurs when the esophagus is exposed to a pH below 4.0.

516. (E) The infant's digestive tract is generally more permeable than that of the adult. This is probably one of the reasons young people are more likely to show allergic reactions. In some animals, this lack of a GI barrier permits the antibodies in the mother's milk to enter the plasma of the young. There is little evidence to indicate that this is the case in human infants, however.

517. (C) The iron deficiency causes a microcytic hypochromic anemia, and the B_{12} deficiency causes a megaloblastic anemia. The acid released by the stomach promotes iron absorption. The parietal cells of the stomach produce a glycoprotein that is needed to transport vitamin B_{12} into the circulation. This so-called intrinsic factor is not needed if the B_{12} is injected intravenously. Removal of the stomach only slightly affects carbohydrate metabolism and absorption in the alimentary tract.

518. (D) The stomach, by storing its contents and slowly releasing them to the intestine, restricts the quantity of hypertonic solution and glucose that enters the more permeable intestine per minute immediately after a meal. In the absence of the stomach, hyperglycemia followed by hypoglycemia and potentially serious decreases in blood volume may occur after a meal. In addition, the amount of food eaten during a single meal is decreased, and the individual must eat more frequently in order to meet his caloric needs. In a normal subject, the absorption of fat from the alimentary tract exceeds 90%. After gastrectomy, it may be less than 30%. This is probably due to the loss of gastric digestion and HCl production and, as a result, a decreased stimulation of pancreatic secretion. There is a decreased secretion of pancreatic juice after gastrectomy.

519. (A) It is through the parasympathetic fibers in the vagi that the brain exerts its major control over the stomach.

520. (E)

521. (E)

CHAPTER 23

Small and Large Intestine

Questions

DIRECTIONS (Questions 522 through 550): Each of the numbered items or incomplete statements in this section is followed by answers or by completions of the statement. Select the ONE lettered answer or completion that is BEST in each case.

522. Which of the following statements is INCORRECT?

(A) the small intestine is about 3 meters long in the healthy subject
(B) the small intestine is connected to the large intestine at the jejunocecal sphincter
(C) the chyme received from the stomach usually remains in the small intestine for 2 hours or longer
(D) usually, over 50% of the digestion and absorption of nutrients is complete by the time the chyme reaches the ileum
(E) glucose absorption from the intestine can occur against a concentration gradient

523. Which of the following statements is INCORRECT?

(A) nicotinic acid, nicotinamide, pantothenic acid, biotin, biocytin, thiamine, thiamine phosphate, folic acid, and flavine mononucleotide are absorbed primarily by active transport
(B) K^+ is actively secreted into the lower half of the small intestine and into the colon
(C) parathormone increases the absorption of Ca^{++} from the small intestine
(D) Fe^{++} enters the mucosa cell of the upper small intestine, where it either combines with apoferritin or moves into the blood
(E) aldosterone exerts a strong influence in facilitating Na^+ absorption in the colon but has a minor influence on the small intestine

524. Vitamin B_{12} deficiency may result from

(A) total gastrectomy
(B) ileal resection
(C) total gastrectomy or ileal resection
(D) inadequate consumption of fresh fruits and leafy green vegetables
(E) high dietary intake of phosphates

525. The upper third of the small intestine has the capacity to actively absorb large quantities of the following EXCEPT

(A) bile salts
(B) pyrimidines
(C) triglycerides
(D) Na^+
(E) neutral amino acids and glucose

526. Which of the following statements is INCORRECT?

(A) a 2-year-old child requires four to eight times the protein intake of a 35-year-old man
(B) after the removal of 60% of the pancreas, over 20% of the ingested protein is lost in the feces
(C) after the removal of the total pancreas, over 40% of the ingested protein is lost in the feces
(D) proteolytic enzymes are released by the chief cells of the stomach, the exocrine cells of the pancreas, and the exocrine cells of the intestine
(E) nucleases released by the cells of the intestine catabolize nucleic acids to pentoses, purine, and pyridimine bases

527. Pancreatic lipase in the intestine catalyzes the conversion of dietary triglycerides to what end products?

(A) diglycerides
(B) 2-monoglycerides
(C) 1-monoglycerides
(D) glycerol
(E) all of the above

528. With regard to the micelle, which of the following is INCORRECT?

(A) a structure with a small polymolecular aggregate containing bile salts, monoglycerides, and fatty acids as its major constituents
(B) one millionth the volume of a fat droplet
(C) frequently seen inside the cells that line the intestine
(D) a structure that speeds lipid absorption
(E) a structure that helps maintain the saturation of the chyme with fatty acids and monoglycerides

529. Which of the following statements is INCORRECT?

(A) most monoglycerides and long-chain (C14 or longer) fatty acids that enter the mucosal cells are re-esterified to triglycerides
(B) most short- and medium-chain fatty acids and some glycerol move from the mucosal cell to the hepatic portal blood
(C) chylomicrons are aggregates of protein, cholesterol, saturated triglycerides, phospholipids, and free fatty acids and are 0.1 to 3.5 μm in diameter
(D) chylomicrons move from the mucosal cell into the lymph capillaries and ducts
(E) chylomicrons seldom remain in the lymph ducts longer than a half hour

530. Which of the following statements is INCORRECT?

(A) the bile acids, cholic and chenodeoxycholic acid, are synthesized in liver cells from cholesterol
(B) bile acids are changed in the liver to deoxycholic and lithocholic acids
(C) many microorganisms in the intestine deconjugate taurocholic and glycocholic acid
(D) bile acids are much less effective detergents than their conjugated salts
(E) bacterial overgrowth in the small intestine can cause a decreased fat absorption

531. Which of the following statements is INCORRECT?

(A) over 90% of the bile salts are absorbed into the blood
(B) under 10% of the bile salt pool is lost each day
(C) removal of the terminal ileum causes an increased synthesis of bile salts by the liver
(D) if over 50% of the bile salts are lost each day, the bile salt pool will decrease
(E) loss to the feces of all or most of the bile salts each day causes a watery diarrhea

532. Which of the following statements is INCORRECT?

(A) an increased excretion of conjugated bilirubin in the urine is caused by ob-

struction of the common bile duct by a gallstone or an increased destruction of erythrocytes
(B) most of the urobilinogen is formed in the intestine
(C) an increased excretion of urobilinogen in the urine can be caused by liver damage
(D) the bile pigments markedly facilitate the absorption of fat
(E) urobilinogen is excreted in the bile

533. Which of the following statements is INCORRECT?

(A) the gallbladder normally has a capacity of less than 60 mL
(B) bile pigments, bile salts, and cholesterol may be 5 to 20 times more concentrated in the healthy gallbladder than in the hepatic duct
(C) bile in the gallbladder has a higher concentration of sodium than in the hepatic duct
(D) in the normal subject, all bile in the hepatic duct passes to the gallbladder
(E) the gallbladder will contract in response to emotional stimuli such as hostility, as well as in response to food in the mouth

534. Which of the following statements is INCORRECT?

(A) cholecystectomy usually causes jaundice
(B) cholecystectomy decreases the bile salt pool by more than 50%
(C) cholecystectomy is followed by a progressive dilation of the bile duct
(D) obstruction of the common bile duct usually increases the prothrombin time
(E) obstruction of the common bile duct usually causes a clay-colored stool

535. Which of the following statements is INCORRECT?

(A) *Escherichia coli,* in the biliary system, will conjugate bilirubin and glucuronic acid and, in so doing, cause the production of gallstones
(B) the calcium salt of bilirubin is less soluble in water than the glucuronic acid conjugate of bilirubin
(C) when the calcium salt of bilirubin precipitates, cholesterol crystals aggregate around it
(D) if the concentration of cholesterol in bile rises above 10%, cholesterol crystals will usually precipitate
(E) if the concentration of bile salts falls below 40%, cholesterol crystals will usually precipitate

536. Which of the following statements is INCORRECT?

(A) cholecystokinin is released from the duodenum in response to fat, egg yolk, and meat
(B) cholecystokinin is carried in the blood to the gallbladder where it facilitates the contraction of the bladder
(C) chewing food can reflexly stimulate parasympathetic neurons in the vagus to initiate the contraction of the gallbladder
(D) hypoglycemia reflexly stimulates vagal parasympathetic neurons to increase bile secretion
(E) vagal stimulation causes the release of an agent, gastrin, which decreases bile secretion

537. Which of the following statements is INCORRECT?

(A) bile salts absorbed from the ileum may decrease the synthesis of additional bile salts but increase the volume of bile produced
(B) secretin causes the production of a bile that has a lower concentration of bile salts and HCO_3^- and a higher pH than bile produced in response to bile salts alone
(C) gastrin increases the volume of bile produced
(D) sympathetic stimulation decreases the production of bile
(E) cholecystokinin increases the production of bile

538. Which of the following statements is correct?

(A) most of the water we drink is absorbed in the stomach
(B) most of the vitamin B_{12} in our diet is absorbed in the stomach and first part of the small intestine
(C) A and B are correct
(D) most of the taurocholate in the alimentary tract is absorbed in the jejunum
(E) most of the calcium is absorbed in the duodenum

539. Which of the following hormones are produced by duodenal cells and increase either the volume or the enzyme content of the exocrine secretion of the pancreas?

(A) secretin
(B) aldosterone
(C) A and B
(D) cholecystokinin (pancreozymin)
(E) A and D

540. Chymotrypsinogen

(A) is secreted primarily by the stomach
(B) is converted to chymotrypsin by the action of trypsin
(C) is converted to chymotrypsin by the action of acid
(D) uses starch as its primary substrate
(E) uses triglycerides as its primary substrate

541. The pH of the duodenum is reduced to 4.0. There follows the secretion of a large volume of pancreatic juice with a high concentration of bicarbonate and a low concentration of enzymes. In addition, gastric secretion decreases and bile secretion increases. Apparently, the acidity in the duodenum has produced these responses by causing the

(A) release of gastrin
(B) release of pancreozymin
(C) release of cholecystokinin
(D) release of secretin
(E) stimulation of adrenergic sympathetic neurons

542. Which of the following statements is INCORRECT?

(A) excess iron in the body causes hemochromatosis (ie, pigmentation of the skin, pancreatic damage and a resultant diabetes, cirrhosis of the liver, gonadal atrophy)
(B) in hemochromatosis, iron absorption decreases but does not cease
(C) in hemochromatosis, most of the excess iron is excreted in the urine
(D) most iron is absorbed in the duodenum and first part of the jejunum
(E) iron is held in the body in the iron-containing proteins ferritin and hemosiderin and is transported in the plasma in combination with the globulin transferrin (siderophilin)

543. Which of the following statements is INCORRECT? In a normal, healthy subject

(A) most of the body's iron is in the hemoglobin molecule
(B) an increase in dietary iron causes little or no increase in iron absorption
(C) ferrous iron (Fe^{++}) is absorbed more readily than ferric iron (Fe^{+++})
(D) an excess of iron in the plasma will be followed by an increased migration of plasma iron into mucosal cells
(E) hemorrhage is followed after an interval of 3 to 4 days by increased iron absorption

544. Which of the following statements is INCORRECT?

(A) decentralization of the intestine causes a disappearance of peristalsis
(B) decentralization and atropinization of the intestine does not cause the disappearance of segmenting movements

(C) distention of the intestine causes the release of serotonin (5-hydroxytryptamine), which increases the frequency and force of the peristaltic contractions
(D) the stimulation of parasympathetic neurons to the intestine increases intestinal secretion and motility
(E) peristaltic waves characteristically travel less than 15 cm and then disappear

545. Which of the following statements is INCORRECT?

(A) antiperistaltic waves occur at almost as great a frequency as peristaltic waves
(B) the stimulation of sympathetic neurons to the intestine causes vasoconstriction, the contraction of the muscularis mucosa, and an increase in the motility of the villi
(C) in the decentralized jejunum, distention and HCl increase the force of peristaltic waves
(D) the absorption of glucose and galactose is inhibited by phlorhizin and other nutrients
(E) the rate of D-glucose absorption is more rapid than that for pentose or D-ribose

546. Which of the following statements is INCORRECT? In human beings, the

(A) ileocecal sphincter, unlike the cardiac and pyloric sphincters, is usually closed
(B) surgical removal of the ileocecal sphincter is followed by the regurgitation of large volumes from the large intestine into the small intestine
(C) ileocecal sphincter is controlled, in part, by the myenteric plexus in the cecum
(D) ileocecal sphincter relaxes when a peristaltic wave in the ileum comes within 2 cm of it
(E) ileocecal sphincter increases its frequency of opening in response to food distending the stomach

547. Which of the following statements is INCORRECT?

(A) the first part of a test meal reaches the cecum in about 4 hours, the splenic flexure in 6 hours, the hepatic flexure in 9 hours, and the sigmoid colon in 12 hours
(B) in defecation, the contractions are frequently sufficient to empty the distal colon as far as the splenic flexure
(C) diarrhea may result from volumes of ileal fluid in excess of 2 liters per day entering the colon
(D) diarrhea may result from magnesium or sulfate salts entering the colon
(E) absence of the myenteric plexus in the descending colon can cause megacolon

548. Which of the following statements is MOST correct? The defecation reflex

(A) is facilitated by food entering the stomach
(B) is eliminated by destruction of the lumbar cord
(C) A and B
(D) is eliminated by the paralysis of skeletal muscle
(E) B and D

549. Which of the following statements is INCORRECT? In the colon

(A) Na^+ and water are removed from the chyme
(B) the concentration of K^+ and HCO_3^- in the chyme is increased
(C) any glucose or amino acids in the chyme are actively absorbed
(D) microorganisms produce vitamins (riboflavin, nicotinic acid, biotin, folic acid, and vitamin K) that will be absorbed into the blood
(E) the amount of flatus produced will more than double if one switches from an average diet to one in which 25% of the caloric intake is pork and beans

550. Which of the following statements is INCORRECT?

(A) more protein in the feces comes from bacteria than from the diet

(B) an increase in the cellulose content in the diet tends to cause an increase in the frequency of bowel movements

(C) constipation is associated with a restlessness, dull headache, loss of appetite, and occasionally nausea, all of which are due to the absorption of toxins from the feces

(D) an individual who defecates fewer than three times a week should not necessarily be considered constipated or unhealthy

(E) ammonium ions are produced in the colon

Answers and Explanations

522. (B) The small intestine empties into the large intestine at the ileocecal sphincter. The jejunum lies between the duodenum and the ileum. The data given in option (A) were obtained on human volunteers by intubation. When the small intestine is removed from the body, it is longer as a result of loss of smooth muscle tone. Glucose in the intestine is absorbed by an active transport system similar to that for glucose reabsorption in the proximal tubule of the renal nephron.

523. (A) Vitamin B_{12} (molecular weight = 1357) is one of the few vitamins that requires an active transport system. Feces contain an average of 75 mEq of potassium per liter, whereas the chyme delivered to the colon contains about 5 mEq of potassium per liter. Feces contain an average of 32 mEq of sodium per liter, whereas the chyme delivered to the colon contains 121 mEq of sodium per liter. A decreased plasma aldosterone concentration in the plasma causes an increased fecal sodium concentration.

524. (C) The absorption of B_{12} is greatly facilitated by its combination with a mucoprotein produced by the stomach (intrinsic factor). Since absorption of the B_{12}–mucoprotein occurs in the terminal ileum, either gastrectomy or ileal resection can cause a B_{12} deficiency. Although fresh fruits and leafy green vegetables are excellent sources of vitamin C, they are inadequate sources of B_{12}. Liver, meat, eggs, and milk are good sources of B_{12}.

525. (A) The upper third of the tract includes the duodenum and jejunum. If the bile salts were absorbed in this area, it would interfere with fat digestion.

526. (B) There is usually sufficient pancreatic reserve so that removal of 60% of the pancreas does not affect either protein or fat catabolism. An adult requires 0.5 to 0.7 g of protein per day. A child (1 to 3 years) requires 4 g per day.

527. (E) Lipid catabolism can be characterized as seen in Figure 23–1.

528. (C) The micelle is an intraluminal structure that speeds absorption of fatty acids and monoglycerides by delivering its contents to the outer surface of the intestinal epithelial cell. Fatty acids and monoglycerides are absorbed chiefly by the epithelial cells of the duodenum and jejunum. Most of the conjugated bile salts, on the other hand, are absorbed in the terminal ileum. The micelle may also contain fat-soluble vitamins and cholesterol. It has a diameter of 4 to 6 μm.

529. (E) Lymph flow is slow, and therefore the nutrients carried by the lymph will be added to the blood over a period of hours. Absorbed fat may remain in the lymph channels for periods up to 12 hours. The chylomicron is 89 to 93% triglyceride, 5 to 9% phospholipid, 1 to 7% free fatty acid, and 0.7 to 1.5% cholesterol by weight. Protein covers

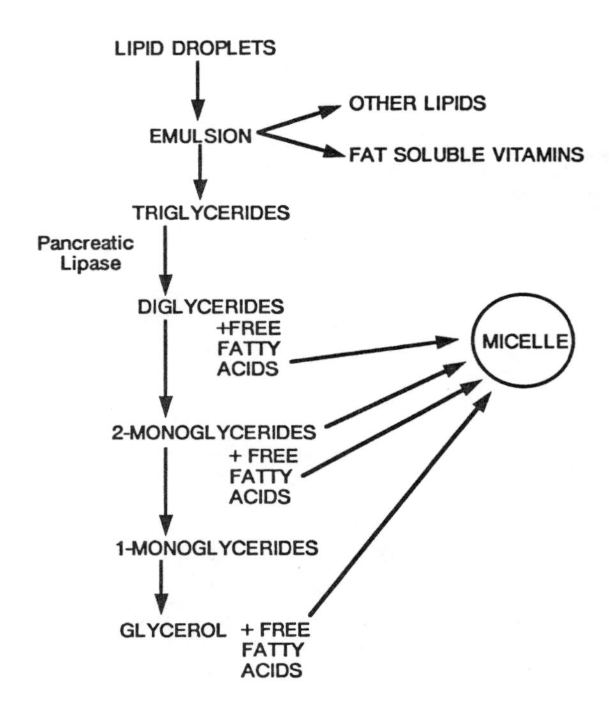

Figure 23–1. Catabolism of lipid by pancreatic lipase.

about 10% of the particle's surface. Chylomicrons are unable to enter blood capillaries. They are able to enter the more permeable lymph capillary.

530. **(B)** Deoxycholic and lithocholic acids are produced in the small or large intestine from the primary bile acids or bile salts by bacterial action. They are sometimes called the secondary bile acids. Cholic acid and chenodeoxycholic acid are conjugated in the liver with taurine and glycine. They are sometimes called the primary bile acids. A decreased fat absorption in the small intestine results from the deconjugation of bile salts by microorganisms.

531. **(B)** Fifteen to 25% of the bile salt pool is lost each day. This is because the pool may circulate twice during the digestion of a single meal. The terminal ileum is where most of the bile salts are absorbed. If over 33% of the pool is lost, the liver is unable to produce enough bile salts to replenish the loss. Normally, 12 to 30 g of bile salts are sent into the duodenum per day. In the absence of bile salt absorption, only 3 to 5 g enter the duodenum per day. Bile salts act as an osmotic cathartic.

532. **(D)** The bile pigments, unlike the bile salts, are waste products of hemoglobin catabolism and play no role in fat absorption. Obstruction of the common bile duct decreases the loss of bile pigments in the feces. An increased destruction of erythrocytes (in hemolytic anemia, for example) will cause an increased production of bilirubin. Urobilinogen is produced in the intestine by bacteria. Most of it is excreted in the feces, but some also diffuses into the blood and is excreted in the urine and bile. After liver damage, the urinary excretion of urobilinogen may increase from 0.5 to 2.0 mg/day to 5 mg/day.

533. **(D)** Bile may move from the hepatic ducts to the common hepatic duct past the cystic duct and into the bile duct and duodenum. In the adult, gallbladder capacity varies between 14 and 60 mL. The gallbladder, by means of the active transport of Na^+, Cl^-, and HCO_3^-, removes water from the bile and modifies the pH of the bile. Its high concentration of sodium results from the sodium being held in association with bile salts in osmotically inactive micelles. The concentration of sodium in gallbladder bile is approximately twice that in hepatic duct bile. The gallbladder, like the stomach, responds to cephalic stimuli.

534. **(A)** After cholecystectomy, there is a dilation of the bile duct, but the major problem remains a reduced storage capacity for bile salts. The maintenance of a low plasma bilirubin concentration by the liver is not a problem because, as bile is produced by the liver, the sphincter of the ampulla is forced open even in the absence of nervous or humoral stimuli. In other words, cholecystectomy impairs the digestive function of the biliary system but not its excretory function.

Obstruction of the common bile duct interferes with both the digestive and excretory functions of the biliary system. By preventing the release of bile salts into the duodenum, more fat is lost in the stool and with it the fat-soluble vitamins A, D, E, and K. Vitamin K is necessary for the production of adequate quantities of prothrombin and other procoagulants. With inadequate vitamin K absorption, the concentration of these procoagulants in the plasma decreases and the prothrombin time increases. The clay-colored stool results in part from a decreased concentration of the brown pigment, urobilinogen, and the increased concentration of lipids in the stool (steatorrhea).

535. (A) *Escherichia coli* have β-glucuronidase activity and tend to hydrolyze the bilirubin conjugate to free bilirubin. When the free bilirubin forms a calcium salt, this salt comes out of solution and forms the nucleus for a gallstone. The formation of conjugates prevents this process. The destruction of the bilirubin conjugate by a microorganism initiates it. Cholesterol is insoluble in water. The bile salts and lecithin form with cholesterol the small, water-soluble aggregates called micelles.

536. (E) The vagus increases the secretion of bile through the action of acetylcholine on liver cells and by stimulating the release of gastrin, which also acts on liver cells. Some important aspects in the control of the gallbladder can be seen in Figure 23–2.

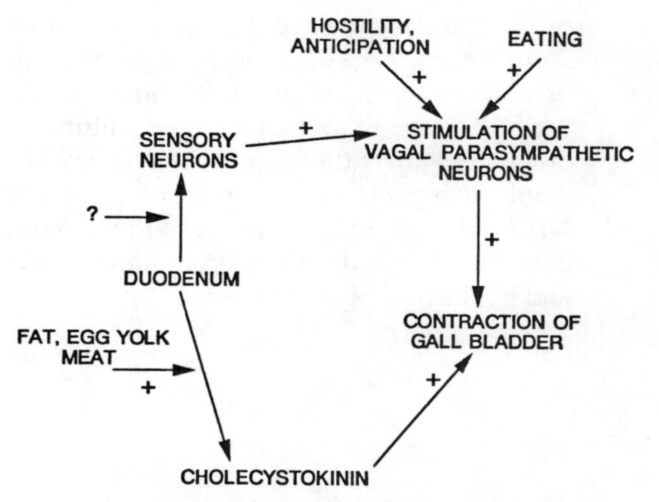

Figure 23–2. Control of the gallbladder.

537. (B) Secretin stimulates the production of bile with a lower concentration of bile salts, a higher concentration of HCO_3^-, and a higher pH than that produced by bile salts alone.

538. (E)

539. (E)

540. (B) The precursor to trypsin (trypsinogen) is also secreted by the pancreas. Chymotrypsinogen is secreted by the pancreatic acinar cells. Trypsin and chymotrypsin reduce polypeptides to smaller peptides.

541. (D) Stimulation of these neurons decreases pancreatic blood flow and tends to decrease secretion. Stimulation of cholinergic neurons to the pancreas, on the other hand, may increase pancreatic secretion. The procedure that will produce the most pronounced increase in pancreatic secretion is the release of secretin. Gastrin facilitates a pancreatic secretion with a high concentration of enzymes and facilitates gastric secretion. Pancreozymin and cholecystokinin are the same hormone. It facilitates the secretion of a pancreatic juice with a high concentration of enzymes and a low concentration of bicarbonate.

542. (C) Most of the excess iron is eliminated in the feces. This is due to (1) the decreased absorption of iron and (2) the sloughing of mucosal cells (desquamation) that have served as a major reservoir of iron.

543. (B) When the diet contains 0.01 mg of iron, approximately 0.001 mg is absorbed. When it contains 10 mg, about 1 mg is absorbed. There are approximately 4 g of iron in the body. Two to 2.5 g are in hemoglobin, 0.5 to 1 g is stored, 150 mg are in myoglobin, 5 mg are in plasma, and the rest is scattered throughout the body in cytochrome, peroxidase, etc. Fe^{++} is absorbed 2 to 15 times more readily than Fe^{+++}. It has been suggested that an interval of 3 to 4 days is, in part, due to the slow migration in the upper intestine of mucosal cells from the area of formation to the tips of the villi.

544. **(A)** The intrinsic nerve plexus is able to continue to coordinate peristalsis in the absence of any connection with the central nervous system. Segmenting movements do not require the presence of any neurons. In humans, unlike rabbits, for example, peristaltic waves passing through long segments of the small intestine (peristaltic rush) are abnormal.

545. **(A)** Reverse peristalsis in human beings is either rare or absent. This does not mean that once chyme has entered the ileum it does not move back into the jejunum. Segmentation, changes in tone, and peristalsis (ie, waves of contraction moving toward the anus) may push some of the chyme toward the stomach as well as in the opposite direction.

Glucose and galactose are actively absorbed from the intestine by a system that has a Tm (transport maximum) and is blocked by phlorhizin. It is because of this active transport that glucose absorption is faster than for many smaller molecules.

546. **(B)** It has been suggested that the major importance of the ileocecal sphincter is (1) to retard emptying of the ileum into the colon and (2) to retard migration of the microorganisms in the colon to the ileum. Options (C), (D), and (E) are statements about the ileocecal sphincter.

547. **(A)** The first part of a test meal reaches the cecum in about 4 hours, the hepatic flexure in 6 hours, the splenic flexure in 9 hours, and the sigmoid colon in 12 hours. The descending colon and sigmoid colon characteristically contain solid material, whereas that in the transverse colon is semisolid and that in the cecum is fluid. In cholera, the delivery of large volumes of isotonic fluid to the lumen of the jejunum and ileum by mucosal cells can cause a loss in the feces of 12 liters of water per day. The salts of magnesium, sulfate, citrate, and tartrate are poorly absorbed in the stomach and intestines and can therefore serve as osmotic cathartics. Megacolon, like achalasia of the esopahagus (enlargement due to food lodging there), is caused by the absence of a myenteric plexus.

548. **(A)** The sacral cord contains the parasympathetic neurons that cause the contraction of the rectum and the relaxation of the internal anal sphincter. The defecation reflex can occur in the absence of skeletal muscle contraction.

549. **(C)** The colon does not actively absorb glucose or amino acids. The value of nutritive enemas is thought to be negligible. Staying on a diet high in beans for 2 weeks has been shown to increase the volume of flatus from 17 to 203 mL/hr.

550. **(C)** Apparently, these symptoms are due to distention of the rectum rather than toxins. What few toxins do enter the blood (NH_4^+, for example) pass via the portal circulation to the liver, where they are removed before they can enter the general circulation. An individual who defecates only once a week, if he does not have the symptoms listed above, should be considered healthy and not constipated. Unfortunately, many confuse what is healthy with what is average. Approximately 10% of the dry weight of the feces is bacteria. In a series of healthy men, it was shown that by changing the fiber content of the diet from 30 to 100 mg/kg/day, the interval between defecations changed from 30 hours to 17 hours. This is not to deny that one's attitudes do not also affect the frequency of defecation. Stool water contains an average of 14 mEq of NH_4^+ per liter. NH_4^+ is produced in the colon by the oxidative deamination of amino acids and by the hydrolysis of urea.

CHAPTER 24

Energy Balance

Questions

DIRECTIONS (Questions 551 through 578): Each of the numbered items or incomplete statements in this section is followed by answers or by completions of the statement. Select the ONE lettered answer or completion that is BEST in each case.

551. Which of the following substances yields less than 4 kcal/mol when its phosphate bond is hydrolyzed?

(A) creatine phosphate
(B) adenosine diphosphate (ADP)
(C) adenosine triphosphate (ATP)
(D) glucose-6-phosphate (G-6-P)
(E) guanosine triphosphate

552. Which of the following statements is INCORRECT?

(A) the anaerobic conversion of 1 mol of glucose to 2 mol of pyruvate results in a net production of 19 mol of ATP from ADP and inorganic phosphate
(B) the aerobic conversion of 1 mol of glucose to CO_2 and H_2O via the Embden–Myerhof pathway and citric acid cycle results in the net production of 38 mol of ATP from ADP and inorganic phosphate
(C) ATP is a precursor to cyclic AMP (cyclic adenosine-3′,5′-monophosphate)
(D) catecholamines ("the first messenger") cause the intracellular release of cyclic AMP ("the second messenger") in the heart, adipose tissue, and liver
(E) the pentoses generated by the hexose-monophosphate shunt are used in the production of nucleotides

553. Which of the following statements is MOST correct? The blood sugar concentration during hypoglycemia is elevated by

(A) the liver, because it has a high concentration of the enzyme glucose-6-phosphatase
(B) skeletal muscle, because it has a high concentration of the enzyme glucose-6-phosphatase
(C) the liver and skeletal muscle, because they have a high concentration of the enzyme glucose-6-phosphatase
(D) the liver and skeletal muscle, because they have a low concentration of the enzyme glucose-6-phosphatase
(E) the liver, because it has a low concentration of the enzyme glucose-6-phosphatase

554. Gluconeogenesis in the liver is decreased by

(A) epinephrine
(B) glucagon
(C) thyroxine
(D) A, B, and C
(E) insulin

555. After a fast of 1 week, the major source(s) of blood glucose is (are)

(A) glycogenolysis
(B) gluconeogenesis from fatty acids
(C) glycogenolysis and gluconeogenesis from fatty acids
(D) glycogenolysis and gluconeogenesis from glycerol
(E) glycogenolysis and gluconeogenesis from amino acids and glycerol

556. Which of the following are ketone bodies?

(A) acetoacetic acid
(B) acetone
(C) A and B
(D) β-OH butyric acid
(E) A, B, and D

557. Which of the following statements is INCORRECT? In a healthy person

(A) the liver produces most of the body's urea
(B) most of the ammonia formed from the deamination of amino acids is changed to uric acid
(C) skeletal muscle degeneration is one cause of an increased concentration of creatinine in the urine
(D) the purines released by the catabolism of nucleotides may be further changed to uric acid
(E) proline is not an essential amino acid because it can be readily produced from glutamic acid as well as other amino acids

558. Patient 1 is on a protein-rich, calorically adequate diet. Patient 2 is on a protein-poor, calorically adequate diet. Patient 1 will have a daily excretion of

(A) urea that is greater than four times that for patient 2
(B) ammonia that is greater than four times that for patient 2
(C) inorganic sulfate that is greater than four times that for patient 2
(D) creatinine that is greater than four times that for patient 2
(E) A and C

559. Which of the following statements is INCORRECT?

(A) free fatty acids in plasma are bound to albumin
(B) unbound free fatty acids in plasma constitute more than 20% of the total free fatty acids
(C) free fatty acids in plasma can be catabolized to CO_2 and water by most tissues
(D) free fatty acids in plasma are in a higher concentration than the ketone bodies in plasma
(E) most of the fatty acids in plasma are in phospholipids and triglycerides and are esterified to cholesterol

560. Which of the following statements is INCORRECT? Insulin

(A) increases the uptake of glucose by cardiac, skeletal, and smooth muscle, adipose tissue, and the liver
(B) has little or no effect on the uptake of glucose by kidney tubules and most of the brain
(C) affects glucose uptake primarily by its action on the cell membrane
(D) has a half-life of about 40 minutes
(E) slows normal growth and wound healing

561. Prolonged fasting and diabetes mellitus lead to

(A) increased gluconeogenesis
(B) ketonemia
(C) a negative nitrogen balance
(D) A, B, and C
(E) hyperglycemia

562. A positive nitrogen balance in the adult occurs

(A) when the dietary intake of protein is increased in the healthy subject
(B) during a debilitating illness
(C) A and B
(D) during recovery from a debilitating illness
(E) during recovery from a debilitating illness or during protein starvation

563. Which of the following produces a markedly more positive nitrogen balance?

(A) testosterone
(B) prolactin
(C) estrogen
(D) chorionic gonadotropin
(E) none of the above

564. Which of the following statements is MOST correct? Growth hormone

(A) decreases the concentration of blood urea nitrogen (BUN)
(B) increases the transport of neutral and basic amino acids into the cell by a mechanism different than that for insulin
(C) has a half-life of 25 minutes
(D) decreases the body's respiratory quotient (RQ)
(E) exerts all of the above actions

565. Which of the following statements BEST describes glucocorticoids, such as cortisol, corticosterone, and cortisone?

(A) they facilitate protein anabolism and closure of the cartilaginous plates of long bones
(B) they facilitate lipid catabolism and cell proliferation in the cartilaginous plates of the long bones of children
(C) they facilitate carbohydrate catabolism and prevent lipid catabolism
(D) they facilitate protein catabolism and hepatic glycogenesis
(E) they act on alpha receptors in blood vessels to cause constriction

566. Which of the following statements is INCORRECT? The cells of the liver normally

(A) produce most of the body's urea and ketone bodies
(B) are the only important site for the production of albumin, prothrombin, and fibrinogen
(C) are the most important site for the production of gamma globulins
(D) inactivate steroid and polypeptide hormones
(E) produce the precursor of angiotensin

567. Which of the following hormones are produced by the thyroid glands?

(A) thyroxine (T_4)
(B) triiodothyronine (T_3)
(C) A and B
(D) calcitonin
(E) A, B, and D

568. Which of the following is NOT a characteristic human response to thyroxine or triiodothyronine?

(A) a decreased O_2 consumption by the adult
(B) little or no change in the O_2 consumption of the brain, testes, lymph nodes, and spleen as compared to the whole adult
(C) an increase in 2,3-diphosphoglycerate in the erythrocyte
(D) an increased nitrogen and calcium excretion in the adult
(E) an increased cholesterol synthesis associated with a movement of cholesterol from the plasma and into the liver

569. Which of the following statements is INCORRECT?

(A) a dietary deficiency of vitamin A and its precursors can cause dry skin and night blindness
(B) a dietary excess of vitamin A can cause a scaly dermatitis and bone pain
(C) a dietary excess of pyridoxine (vitamin B_6) causes hypoirritability and lethargy
(D) a dietary deficiency of copper or iron can cause abnormalities in the blood
(E) dietary magnesium, manganese, cobalt, bromine, and zinc are necessary in trace amounts

570. An individual on a diet that is practically devoid of cholesterol will in many cases have

(A) a low blood cholesterol level because the body cannot synthesize cholesterol
(B) a near normal blood cholesterol level because the liver synthesizes cholesterol and adds it to the blood when the blood level is low
(C) a near normal blood cholesterol level because the intestinal flora supply cholesterol to the blood
(D) a near normal blood cholesterol because of an increased ability of the intestine to absorb cholesterol
(E) a near normal blood cholesterol because of a decreased excretion of cholesterol in the bile

571. A subject walking 5 miles/hr at a 5% slope on a treadmill had an oxygen consumption of 2 L/min (STPD) and a respiratory quotient (RQ) of 0.8. If his caloric equivalent for oxygen was 4.8 kcal/L, approximately how much heat must he eliminate from his body in order to maintain a constant body temperature?

(A) less than 2 kcal/min
(B) between 2 and 4 kcal/min
(C) between 4 and 6 kcal/min
(D) between 6 and 8 kcal/min
(E) over 8 kcal/min

572. If, in the exercise study in Question 571, the subject had an RQ of 0.75, this lower RQ would

(A) cause a decrease in the caloric equivalent for 1 liter of O_2
(B) indicate that lipid catabolism is a less important source of energy than protein or carbohydrate catabolism
(C) indicate that both of the above statements are true
(D) cause a decrease in alveolar minute ventilation
(E) A, B, and D

573. Which of the following statements is correct? In a resting individual in the postabsorptive state, the RQ of

(A) practically all major organs (heart, brain, skeletal muscle, etc.) is similar
(B) the brain is greater than 0.95
(C) skeletal muscle is greater than 0.95
(D) the heart is greater than 0.95
(E) the skin is greater than 0.95

574. Which of the following statements is INCORRECT? The specific dynamic action of

(A) food causes an increase in a subject's O_2 consumption after eating
(B) food persists for about 6 hours after eating
(C) protein is due in large part to the oxidative deamination of amino acids in the liver
(D) protein is in excess of 20% of its total caloric value
(E) fat is greater than that for protein

575. Which of the following statements is INCORRECT?

(A) the kcal (kilocalorie) is the amount of heat required to raise the temperature of 1 g of water 1°C (eg, from 15 to 16°C)
(B) the basal metabolic rate of an average 40-year-old man is about 1600 kcal per 24 hours

(C) the total metabolic rate of an average 40-year-old man is about 3000 kcal per 24 hours
(D) the total metabolic rate is increased following eating
(E) when one changes the room temperature from 30 to 0°C or from 30 to 50°C, the total metabolic rate increases

576. Which of the following statements is INCORRECT?

(A) leafy green vegetables are good sources for folic acid, vitamin C, and vitamin K
(B) liver is a good source for thiamine, riboflavin, niacin, pantothenic acid, biotin, and vitamin B_{12}
(C) 1 g of fat produces more than twice the number of calories as 1 g of carbohydrate
(D) foods, such as margarine, that are derived from vegetables have a higher cholesterol concentration than most foods derived from animals, such as milk, cheese, eggs, and meat
(E) estrogens cause a lowering of plasma cholesterol levels

577. Which of the following statements is MOST accurate and complete?

(A) the major feeding and satiety centers lie in the brain stem
(B) the high blood sugar that occurs during diabetes mellitus depresses the hunger center
(C) insulin facilitates the utilization of sugar by the "glucostats" in the brain
(D) appetite is unaffected if the stomach is distended with inert material
(E) all of the above

578. Which of the following statements is INCORRECT? Thirst is

(A) not quenched when water is drunk but when the water is absorbed into the blood
(B) diminished by lesions in the hypothalamus that do not affect food intake
(C) increased by infusing hypertonic solutions into the arteries going to the hypothalamus
(D) produced by a fall in blood volume in which there is no change in blood tonicity
(E) increased in diabetes mellitus

Answers and Explanations

551. (D) G-6-P yields 2 to 3 kcal/mol when its phosphate bond is hydrolyzed. This bond is not considered a high-energy phosphate bond. Creatine phosphate, ADP, ATP, and guanosine triphosphate yield 10 to 12 kcal/mol when their terminal phosphate bond is hydrolyzed.

552. (A) In glycolysis, there is a net production of 2 mol of ATP. The cyclic AMP released in the heart increases the force of contraction, in adipose tissue causes lipolysis, and in the liver causes glycogenolysis.

553. (A) Reactions in liver and kidney are seen in Figure 24–1. Skeletal muscle can release large concentrations of lactic acid into the blood but no glucose. This is because it lacks the enzyme (glucose-6-phosphatase) for converting glucose-6-phosphate (G-6-P) to glucose. It can, however, perform the reactions seen in Figure 24–2.

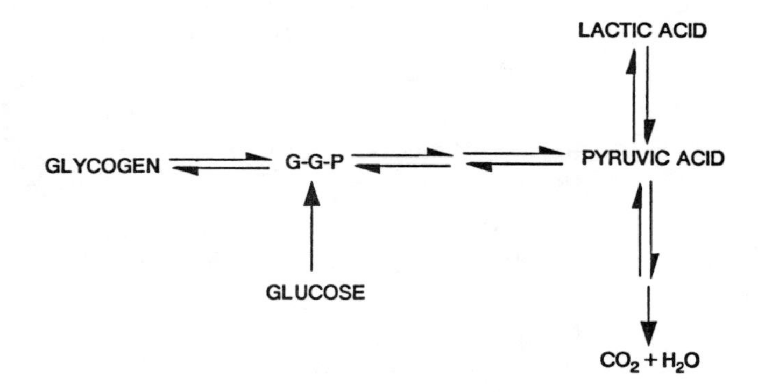

Figure 24–2. Some glycolytic reactions occurring in skeletal muscle.

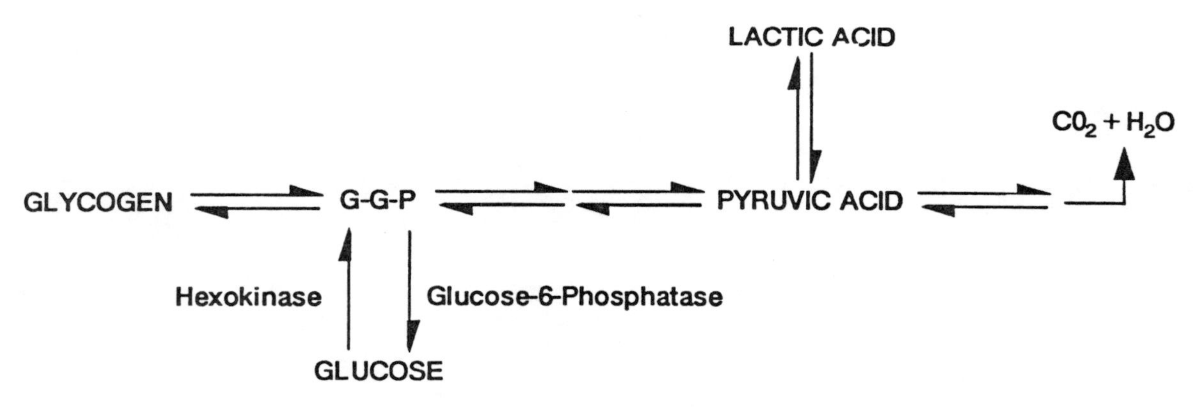

Figure 24–1. Some glycolytic reactions occurring in liver and kidney.

554. (E) Insulin inhibits hepatic enzymes that facilitate the catabolism of proteins and fats, facilitates the production of glycogen and fatty acids from glucose-6-phosphate, and lowers the blood sugar. Epinephrine and glucagon antagonize these actions.

555. (E) Glycerol and the gluconeogenic amino acids (eg, cystine, glycine, serine, valine, threonine) form glucose. The so-called ketogenic amino acids include leucine, isoleucine, phenylalanine, and tyrosine. After 48 hours of fasting, most of the body's glycogen has been depleted. Fatty acids do not form glucose in important quantities.

556. (E) A ketone body is a molecule that contains a C=O group (ie, a ketone group) and is produced in human beings and other vertebrates in the liver from acetyl-CoA. The ketone bodies are (1) acetoacetic acid, (2) β-OH butyric acid, and (3) acetone.

557. (B) Most of the ammonia produced in human beings is changed to urea. Small amounts are used to produce uric acid. Glutamine serves to transport nitrogen released during deamination to the kidney, where it may be excreted as NH_4^+, or to the liver, where it may be converted to urea. The liver and brain are probably the only two organs in the body that produce urea. Creatinine is formed by the removal of phosphate from creatine-phosphate. Urine contains about 42 mEq of uric acid and 1820 mEq of urea per liter. The essential amino acids include (1) valine, (2) leucine, (3) isoleucine, (4) threonine, (5) methionine, (6) phenylalanine, (7) tryptophan, and (8) lysine.

558. (E) Excess protein intake causes increased gluconeogenesis and the increased excretion of the nitrogenous waste product, urea, and of inorganic sulfate. On the other hand, the quantity of ammonia and creatinine excreted is determined not by protein ingestion, but by the pH of the plasma and the "wear and tear" on the tissues, respectively. The amount of uric acid excretion is related to both protein ingestion and "wear and tear."

559. (B) Practically all free fatty acids in plasma are bound to albumin. In statement (C), the major exception to this rule is the brain, which depends on carbohydrate for nutrition. The heart and skeletal muscle, on the other hand, usually receive most of their energy from the catabolism of fatty acids. Ketone bodies, in the healthy resting subject, have a plasma concentration of 1 mg/100 mL, whereas that for free fatty acids is 12 mg/100 mL. Approximately 50% of the total fatty acids in plasma are in phospholipids, 25% in triglycerides, 20% esterified to cholesterol, and 5% are free (ie, loosely bound to plasma proteins).

560. (E) Insulin increases the uptake of amino acids by the cell and increases protein synthesis by the cell.

561. (D) Hyperglycemia is characteristic of diabetes mellitus, but in starvation there is a hypoglycemia. The ketonemia in both conditions is due to an increased production of ketone bodies by the liver. The negative nitrogen balance is due to an increased catabolism of proteins by the cells and a resultant increased production of urea and other nitrogen-containing waste products.

562. (D) Growth and repair of body tissues causes a positive nitrogen balance. An increased intake of dietary protein does not change the nitrogen balance in the healthy subject. The excess protein in the diet is converted to other nutrients (fat and carbohydrate), and the nitrogen in the protein molecules is transferred to nitrogen-containing waste products such as urea. During a debilitating illness, the nitrogen excretion usually exceeds its intake and there is a negative nitrogen balance. During protein starvation, the excretion of nitrogen in the urine exceeds the intake in the food.

563. (A) Testosterone increases protein synthesis.

564. (E) Growth hormone (or somatotrophic hormone [STH]) facilitates protein synthesis. A longer half-life would be inconsistent with growth hormone's role of the minute-to-minute control of body metabolism. Growth hormone decreases carbohydrate catabolism and increases fat catabolism.

565. **(D)**

566. **(C)** Although 60 to 80% of the plasma globulins are produced in the liver, the plasma cells in the lymphatic system are probably the most important source of gamma globulin. Cortisol, aldosterone, testosterone, estrogens, insulin, progesterone, growth hormone, prolactin, antidiuretic hormone, and oxytocin are all changed in the liver to less active substances. The liver produces α-2-globulin.

567. **(E)**

568. **(A)** Thyroxine and triiodothyronine can increase the O_2 consumption of human beings by more than 50% while producing less than a 5% change in the O_2 consumption of the brain. The protein catabolism may be so marked in skeletal muscle as to cause weakness and creatinuria. The action of T_4 and T_3 on bone may cause a mild osteoporosis and hypercalciuria.

569. **(C)** Hypovitaminosis B_6 causes hyperirritability and convulsions. Hypervitaminosis symptoms for this water-soluble vitamin apparently do not occur. The fat-soluble vitamins A and D, unlike the water-soluble vitamins, cause symptoms of hypervitaminosis. Copper deficiency causes neutropenia, and iron deficiency causes anemia. Although essential in trace amounts, magnesium, manganese, cobalt, bromine, and zinc are seldom, if ever, deficient in diets that are otherwise normal.

570. **(B)** Most human tissues can synthesize cholesterol. The liver is the most important organ for the maintenance of blood cholesterol levels. Although a lower blood cholesterol may result from dietary restriction, it is not because the body cannot synthesize it. In human beings, the microorganisms in the intestine probably serve only as a source of certain vitamins (possibly only folic acid in the adult). A change in the absorptive power of the intestine for Ca^{++} may serve to increase the level of blood Ca^{++}. Changes in the absorptive power for cholesterol, on the other hand, are not an important response to hypocholesterolemia. The cholesterol in the bile is reabsorbed into the blood in the intestinal ileum. In short, cholesterol in the bile is not an important route for loss of cholesterol from the body. The bile pigments, on the other hand, are bile elements lost from the body in the feces of the normal subject.

571. **(E)**

$$\text{Energy liberated} = 2 \text{ L of } O_2/\text{min}) \times (4.8 \text{ kcal/L of } O_2) = 9.6 \text{ kcal/min}$$

572. **(A)** RQ = CO_2 produced/O_2 utilized. Therefore, a decrease in RQ means it takes more O_2 to produce the same quantity of CO_2 (ie, approximately the same number of calories). Compared to carbohydrate and protein, lipid is oxygen poor in terms of content. A decrease in lipid catabolism would increase RQ (ie, decrease the O_2 utilized). Subjects with a low RQ have a greater O_2 consumption. A greater O_2 consumption is usually associated with an increased alveolar minute volume.

573. **(B)** The RQ averages 0.83 for an adult, 0.98 for the brain, and 0.85 for skeletal muscle, heart, and skin.

574. **(E)** The specific dynamic action of protein is 30% of its caloric value, whereas that for carbohydrate and fat is about 6% and 4%, respectively.

575. **(A)** 1 dyne^{-cm} = 1 erg = 10^{-7} joules = 2.39×10^{-8} cal = 2.39×10^{-11} kcal. The kcal is the amount of heat required to raise the temperature of 1 kg of water from 15 to 16° C. For a woman, the basal metabolic rate is about 1300, and for an 18-year-old boy about 1800 kcal. The total metabolic rate is increased following eating due to the specific dynamic action of food. Increases in total metabolic rate occur in response to body cooling (may cause shivering) or body warming (may cause sweating).

576. (D) Cholesterol is found only in animals and their products. One gram of animal fat yields about 9.5 kcal, whereas 1 g of starch yields 4.2 kcal, 1 g of cane sugar yields 4.0 kcal, and 1 g of protein yields 4.4 kcal. Atherosclerosis is characterized as a condition in which cholesterol infiltrates arterial walls. It has been suggested that estrogens, in that they lower plasma cholesterol levels, protect premenopausal women from atherosclerosis.

577. (C) Appetite is controlled, in part, by the utilization of blood sugar (ie, arteriovenous glucose difference) in cells that are said to constitute the glucostat. The feeding center is in the nuclei of the lateral part of the hypothalamus and the satiety center is in the medial hypothalamic nuclei. The amygdaloid nuclei of the limbic system also play a role in controlling appetite. Distention of the stomach decreases appetite.

578. (A) Water in the mouth and stomach temporarily inhibits thirst by the stimulation of receptors and their sensory neurons.

CHAPTER 25

Body Temperature
Questions

DIRECTIONS (Questions 579 through 593): Each of the numbered items or incomplete statements in this section is followed by answers or by completions of the statement. Select the ONE lettered answer or completion that is BEST in each case.

579. Mechanisms for the elimination of heat from the body are

(A) evaporation
(B) convection
(C) A and B
(D) radiation
(E) A, B, and D

580. A 25-year-old man clothed only in shoes and shorts is in a room with a temperature of 70°F (21°C) and a humidity of 20%. As he stands quietly, what percentage of the total heat that he loses from his body will be dissipated by the mechanisms listed in Question 579?

(A) 65% by insensible perspiration
(B) 65% by radiation
(C) 40% by vaporization and 40% by radiation
(D) approximately equal amounts by evaporation, radiation, and convection
(E) approximately equal amounts by evaporation, radiation, convection, and conduction

581. The subject in Question 580 would, during a strenuous tennis match, lose most of his body heat by

(A) conduction
(B) convection
(C) radiation
(D) evaporation
(E) no single mechanism

582. If the subject in Question 580 is placed in a room with a temperature of 37°C and a humidity of 20%,

(A) over 90% of the heat lost will be by evaporation
(B) most of the heat lost will be by evaporation
(C) most of the heat lost will be by radiation
(D) most of the heat lost will be by convection
(E) most of the heat lost will be by conduction

583. Which of the following statements is correct for a healthy, naked, 24-year-old woman under resting conditions?

(A) her rectal temperature averages 37°C (98.6°F)
(B) the rectal temperature has a diurnal fluctuation of about 0.6°C, being lowest in most individuals at about 6 A.M.
(C) A and B
(D) during the menstrual cycle, the most distinct rise in rectal and oral temperature occurs about 2 days before menstruation
(E) A, B, and D

584. Which of the following statements is correct?

(A) during strenuous exercise, the rectal temperature may rise to 40°C
(B) during exercise, the skin temperature may either rise (frequently to 39°C) or fall
(C) for each rise of 1°C in fever, there is a 13% increase in O_2 consumption
(D) rectal temperatures above 43°C may be fatal
(E) all are correct

585. Which of the following statements is correct?

(A) human subjects with rectal temperatures between 21 and 24°C for 6 hours can be revived if warmed rapidly
(B) at a rectal temperature of 24°C, the subject is conscious and shivering
(C) at a rectal temperature of 24°C, the heart rate and respiratory rate are slightly elevated
(D) at a rectal temperature of 24°C, hemorrhagic problems are exaggerated
(E) an individual usually tolerates well a rectal temperature of 32°C maintained for 2 days under hospital conditions

586. The countercurrent mechanism of heat transfer during exposure to the cold explains which of the following sets of data?

(A) heat in arteries of the arm is 37°C, in arteries of the hand is 22°C, in veins of the arm is 36°C
(B) heat in arteries of the arm is 37°C, in arteries of the hand is 36°C, in veins of the arm is 35°C
(C) heat in arteries of the arm is 35°C, in arteries of the hand is 36°C, in veins of the arm is 37°C
(D) heat in arteries of the arm is 36°C, in arteries of the hand is 22°C, in veins of the arm is 37°C
(E) heat in arteries of the arm is 33°C, in arteries of the hand is 25°C, in veins of the arm is 22°C

587. Which of the following statements is INCORRECT? Under resting conditions at a temperature of 75°C,

(A) a naked black man absorbs 20% more heat from the sun than a naked white man of similar height and weight
(B) approximately one third of the heat lost due to insensible perspiration is lost in the expired air
(C) 580 kcal of heat are lost from the body for each kg of water vaporized on the skin or mucous membranes
(D) sweat glands are innervated by postganglionic, cholinergic, sympathetic neurons
(E) an increased cutaneous blood flow causes an increased thermal conductivity of the skin

588. An athlete ran a 42-km race in 158 minutes. During the race, he lost 3 liters of sweat. After the race, he drank 3 liters of water. Several hours later, the subject was studied. Which of the following conclusions would be MOST likely? Because sweat contains

(A) a higher concentration of Na than the extracellular fluid, there was hyponatremia
(B) a lower concentration of Na than the extracellular fluid, there was hypernatremia
(C) approximately the same concentration of Na as the extracellular fluid and there was water replacement but not Na replacement, there was an increase in blood volume
(D) approximately the same concentration of Na as the extracellular fluid and there was water replacement but not Na replacement, there was a decrease in blood volume
(E) a lower concentration of Na than the extracellular fluid and there was water replacement but not Na replacement, there was a decrease in blood volume

589. A patient receives a spinal transection at T2, and after a period of time, his withdrawal reflexes return below the lesion. If his legs and thighs are now cooled, the initial response to this cooling would be

(A) shivering in the legs because of a direct action of cooling on the legs
(B) shivering in the legs because of a spinal reflex
(C) shivering in the arms because of a spinal reflex
(D) shivering in the legs because of a cooling of the hypothalamus
(E) shivering in the arms because of a cooling of the hypothalamus

590. A patient with a spinal transection at T8 differs from one with a transection at T1 in that the former responds to

(A) cooling of the lower extremities with a more intense vasoconstriction in the hand
(B) warming of the lower extremities with a more profuse sweating in the hand
(C) A and B
(D) cooling of the lower extremities with a more intense vasodilation in the hand
(E) B and D

591. A group of young men living in a cold environment began exercising for 5 hours at 36°C. After 12 consecutive days of this regime, what signs of acclimatization to exercising at this elevated temperature might they show?

(A) the volume of sweat produced during exercise would approximately double
(B) they would have less marked increases in rectal temperature during exercise
(C) A and B
(D) the volume of sweat produced during exercise would decrease to approximately one half
(E) the O_2 used during exercise would increase

592. An individual exposed to a hot environment will have an increased body core temperature, which will cause

(A) a reflex stimulation of cholinergic sympathetic neurons to the sweat glands
(B) a reflex inhibition of adrenergic sympathetic neurons to the blood vessels of the skin
(C) A and B
(D) a decreased skin temperature
(E) A, B, and D

593. A fever induced by a bacterial toxin is caused by

(A) the production of an endogenous pyrogen by the leukocytes and macrophages
(B) a direct action of the toxin on skeletal muscle and the cutaneous blood vessels
(C) a direct action of the toxin on the hypothalamus
(D) an elevation of the set point in the heat regulatory center of the substantia nigra
(E) an inhibition of prostaglandin synthesis and release in the forebrain

Answers and Explanations

579. (E) Evaporation is the change of a solid or liquid (water in this case) into a gas (water vapor). Convection is the transmission of heat in liquids or gases by fluid moving over and away from a structure (ie, air moving over the skin). Radiation is the transmission of energy in all directions to distant objects. Another means is conduction, which is the transmission of energy (such as heat) between adjacent structures without the movement of either structure.

580. (B) Radiation accounts for about 65% of the heat dissipated, evaporation 20%, and convection about 15%.

581. (D) During strenuous exercise, the amount of heat lost by radiation may increase by 10%, by convection may triple, and by evaporation may increase 25-fold.

582. (A) Radiation, convection, and conduction of heat from the body require a room temperature lower than the temperature of the skin. At a room temperature of 36°C or higher, all of the heat lost from the body is usually by evaporation. This requires, however, a humidity below 100%.

583. (E) The most distinct rise occurs after ovulation. The rectal temperature returns to the preovulation value several days prior to menstruation. At a room temperature of 23°C, the average skin temperature will be 5°C below rectal temperature and the foot will be over 11°C below rectal temperature.

584. (E) During most exercises, there is an increased blood supply to the skin. This is an important mechanism for shunting heat from the core to the shell. During a maximal effort, however, cutaneous blood flow may go to 20% of the pre-exercise value. Under these circumstances, the rectal temperature rises rapidly and the cutaneous temperature falls. The temperature of the skin will also depend on the quantity and type of clothing, the amount of sweating, the humidity, the room temperature, and the amount of air movement. If a rectal temperature above 43°C is maintained for an hour, it is usually fatal. There is, however, individual variability.

585. (E) This level of hypothermia slows the heart and decreases body metabolism. Prolonged cooling at 21 to 24°C causes cardiac standstill or fibrillation; revival is possible only for short periods (eg, 10 minutes). At rectal temperatures below 28°C, the subject is unconscious, not shivering, and unable to recover from cooling by his own mechanisms. At a rectal temperature of 24°C, the heart rate, arterial pressure, and respiratory rate are reduced, and bleeding is a reduced problem.

586. (A) During exposure to the cold, blood flow to the hand is reduced, and much of the heat in the arterial blood in the arteries of the arm and forearm is conducted to the cooler veins that lie adjacent to these arteries (the venae comites). Thus, the veins and arteries of the arm have blood with a relatively high temperature, and the blood in the hand has a low temperature. This countercurrent heat mechanism is similar to that responsible for Na^+ transfer in the ascending and descending loops of the nephron and is an important mechanism for heat conservation.

587. (A) Human skin, as far as heat is concerned, can be characterized as a 97% black body. In other words, the skins of blacks and whites, although different in the visual ranges, are similar in their heat absorption characteristics. The skins of blacks, on the other hand, let less ultraviolet light (wavelengths 290 to 32 μm) penetrate and are probably therefore less susceptible to hypervitaminosis D and skin cancer. In option (C), of this 580 kcal, 539 kcal is the heat of vaporization and 41 kcal is the heat carried from the body by the vapor.

588. (E) Loss of Na plus gain of water leads to hypotonicity of the extracellular compartment. Diffusion of water into the intracellular environment then decreases the blood volume. Sweat losses as great as 10 liters per day have been reported. Sweat is hypotonic and contains from 9 to 80 mEq of Na per liter. Perfuse sweating without water or Na replacement tends to produce hypernatremia but with water replacement may cause hyponatremia. The adrenal cortex responds to hypernatremia by decreasing its output of aldosterone. There results an increased concentration of Na in the urine and sweat.

589. (E) By cooling the lower extremities, the blood (and therefore the hypothalamus) has been cooled. In option (D), the transection has eliminated functional connections between the hypothalamus and the legs.

590. (C) A transection at T8, unlike the one at T1, does not separate the hypothalamus from the adrenergic sympathetic neurons to the arterioles of the hand nor the cholinergic sympathetic neurons to the sweat glands of the hand.

591. (C) The volume of sweat produced during exercise would approximately double (ie, change from 600 mL/hr during the first hour of exercise on the first day to 1200 mL/hr during the first hour on the twelfth day). The concentration of Na in the sweat would decrease. With a more effective means of eliminating excess heat (sweating and circulatory changes) and an increased efficiency, the subject would have less marked increases in rectal temperature during exercise. The O_2 used during exercise would decrease.

592. (C) The decreased adrenergic sympathetic tone to the cutaneous vessels shunts more blood to the skin. This causes an increased skin temperature and an increased heat loss from the body if the temperature of the external environment is lower than body temperature. Evaporation on the skin surface cools the skin and therefore the body, even when the external temperature exceeds body temperature.

593. (A) Polymorphonuclear leukocytes, monocytes, splenic and alveolar macrophages, and Kupffer cells produce the endogenous pyrogen in response to the toxin. The endogenous pyrogen elevates the set point of the heat regulatory centers in the preoptic area of the hypothalamus (possibly by stimulating the local release of prostaglandins).

PART VIII

Endocrine Control

CHAPTER 26

Hormones

Questions

DIRECTIONS (Questions 594 through 602): Each set of items in this section consists of a list of lettered options followed by several numbered words or phrases. For each numbered word or phrase, select the ONE lettered option that is most closely associated with it. Each lettered option may be selected once, more than once, or not at all.

Questions 594 through 599

Match the hormones to the list of structural formulas in Figure 26–1.

594. Epinephrine

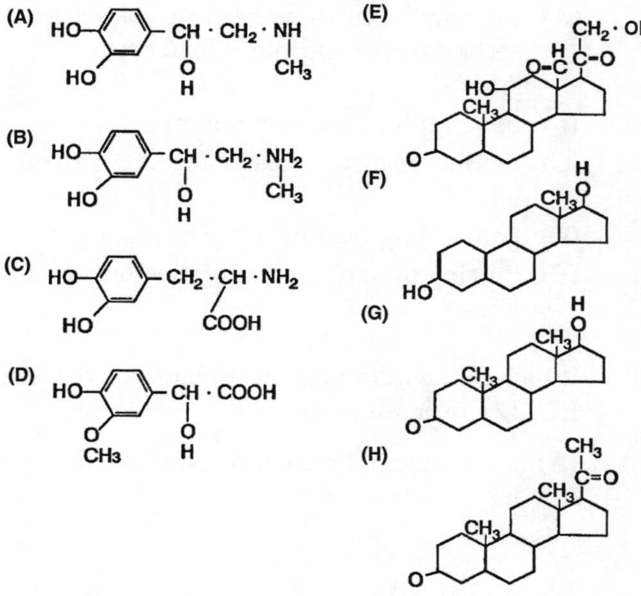

(A) HO, HO, CH · CH_2 · NH, O, H, CH_3

(B) HO, HO, CH · CH_2 · NH_2, O, H, CH_3

(C) HO, HO, CH_2 · CH · NH_2, COOH

(D) HO, O, CH_3, CH · COOH, O, H

(E) CH_2 · OH, H, C–O, O–C, HO, CH_3, O

(F) H, O, CH_3, HO

(G) H, O, CH_3, CH_3, O

(H) CH_3, C=O, CH_3, CH_3, O

Figure 26–1. Structural formulas of various compounds.

595. Norepinephrine

596. Direct precursor of norepinephrine synthesis

597. Estradiol

598. Testosterone

599. Progesterone

Questions 600 through 602

Match the following list of hormones with their role in blood sugar control (see Fig. 26–2).

(A) insulin
(B) glucagon
(C) C-peptide
(D) epinephrine
(E) norepinephrine

600. Hormone X

601. Hormone Y

602. Hormone Z

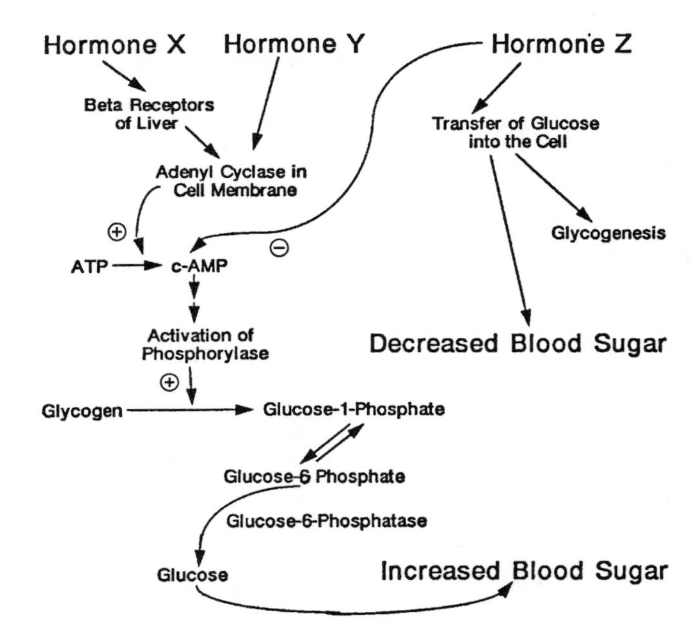

Figure 26–2. Control of blood sugar level.

DIRECTIONS (Questions 603 through 609): Each of the numbered items or incomplete statements in this section is followed by answers or by completions of the statement. Select the ONE lettered answer or completion that is BEST in each case.

603. The 17-hydroxycorticoids include

(A) 11-desoxycortisol, cortisol, and cortisone
(B) progesterone and aldosterone
(C) estradiol and testosterone
(D) progesterone, aldosterone, 11-desoxy-cortisol, cortisol, and cortisone
(E) only cortisol

604. The 17-ketosteroids in the urine include all of the following EXCEPT

(A) androgens produced by the adrenal cortex
(B) testosterone
(C) metabolic products of testosterone
(D) metabolic products of cortisone
(E) metabolic products of cortisol

605. Prostaglandins

(A) are produced only in the prostate gland
(B) have, as their major source, the prostate gland
(C) are produced by most of the organs in the body
(D) produce biologic changes only at molar concentrations ten times those necessary for most well-established hormones
(E) have little or no known action on non-contractile cells

606. Which of the following statements is INCORRECT?

(A) adenylate cyclase is found in the cell membrane of liver cells
(B) adenylate cyclase is activated in the liver cell in response to epinephrine
(C) adenylate cyclase catalyzes the conversion of adenosine triphosphate (ATP) to cyclic adenosine monophosphate (cAMP)
(D) cAMP inhibits phosphorylase in the liver
(E) cAMP increases the mobilization of Ca^{++} from bone (ie, increases blood Ca^{++})

607. Steroid hormones control cell function by all of the following EXCEPT

(A) causing the cell membrane to release a second messenger other than cyclic AMP
(B) penetrating the cell membrane
(C) binding to a cytoplasmic receptor protein
(D) penetrating the nuclear membrane
(E) causing the production of messenger RNA

608. Which of the following statements is INCORRECT? Corticosteroids

(A) in excess can produce a condition similar to diabetes mellitus

(B) decrease the vasoconstrictor action of catecholamines
(C) play a permissive role in the response of adipose tissue to growth hormone
(D) facilitate the release of free fatty acids into the blood by adipose tissue
(E) increase blood sugar levels

609. Which of the following statements is INCORRECT? In Graves' disease there is usually

(A) an enlargement of the thyroid gland
(B) a release of LATS, which is an immunoglobulin
(C) a release of LATS, which causes an increased release of TSH
(D) a release of LATS, which causes an increased release of thyroid hormone
(E) a release of LATS, which, unlike TSH, crosses the placental barrier

Answers and Explanations

594–599. (594-A, 595-B, 596-C, 597-F, 598-G, 599-H) (C) is dopamine, and (E) is aldosterone.

600–602. (600-D, 601-B, 602-A) Epinephrine and glucagon promote glycogenolysis and glucose release by activating live cell adenyl cyclase. Insulin accelerates glucose transfer into cells by enhancing glucose phosphorylation.

603. (A) The C-21 steroids with a hydroxyl group in the C-17 position are called 17-hydroxycorticoids. The C-19 steroids with an oxygen in the C-17 position are called 17-ketosteroids and include dehydroepiandrosterone, androsterone, and androstenedione.

604. (B) The androgens produced by the adrenal cortex and testis are 17-ketosteroids. Testosterone, on the other hand, is a 17-hydroxycorticoid. In addition, about 10% of the cortisol secreted (this includes that which is later changed to cortisone) is converted in the liver to a 17-ketosteroid. In total, the normal adult man eliminates in the urine about 15 mg of 17-ketosteroids per day. About two thirds of these come from the adrenal cortex and one third from the testis. The normal adult woman, on the other hand, eliminates about 10 mg/day in her urine.

605. (C) In 1930, Kurzok and Lieb noted that fresh human semen caused uterine strips to either contract or relax. U.S. von Euler, in Sweden, called the active substance prostaglandin. It is now believed that most of the prostaglandin in the semen comes from the seminiferous tubules and not the prostate. The kidneys, GI tract, hypothalamus, and many other cells also make prostaglandins. Prostaglandins are active at concentrations of 10^{-9} molar. This means that they are probably more potent than any of the well-established hormones. PGE (1) produces vasodilation, bronchodilation, and contraction of the uterus, (2) produces secretion of the adrenal cortex, corpus luteum, and thyroid, (3) produces natriuresis, and (4) inhibits lipolysis, platelet aggregation, and the secretion of acid by the stomach.

606. (D) Cyclic AMP activates phosphorylase in the liver.

607. (A) The steroids function usually through (1) their penetration of the cell membrane, (2) their formation of a protein complex in the cytoplasm, (3) the entry of the complex into the karyoplasm, (4) its facilitation of the formation of messenger RNA by DNA, (5) the facilitation of protein synthesis in the cytoplasm by mRNA. Thyroid hormone also penetrates the cell membrane and increases the production of mRNA.

608. (B) Corticosteroids play a permissive role in the response of a number of cells to glucagon, epinephrine, and norepinephrine. They inhibit the enzyme catecholamine-O-methyl-transferase. They decrease the sensitivity of cells to insulin. They facilitate gluconeogenesis by causing lipid, protein, and amino acid catabolism and inhibiting the action of insulin.

609. (C) Graves' disease is a form of hyperthyroidism caused by a group of antibodies against TSH (thyroid-stimulating hormone) receptors in the thyroid. These antibodies (thyroid-stimulating immunoglobulins [TSI]) include a long-acting thyroid stimulator (LATS). They are produced by lymphocytes and have an effect on the thyroid similar to that produced by TSH but, unlike TSH, their production and secretion are not inhibited by thyroid hormones. Since LATS causes an increased plasma concentration of thyroid hormones, it will cause a decreased release of TSH (see Fig. 26–3).

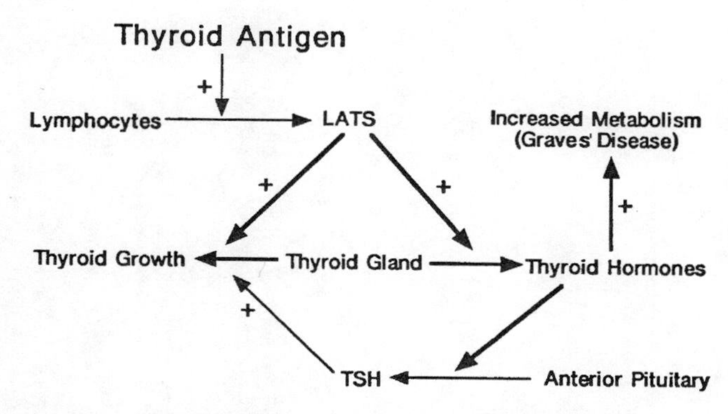

Figure 26–3. Role of LATS in Graves' disease.

CHAPTER 27

Thyroid and Parathyroid Glands

Questions

DIRECTIONS (Questions 610 through 622): Each of the numbered items or incomplete statements in this section is followed by answers or by completions of the statement. Select the ONE lettered answer or completion that is BEST in each case.

610. 3,5,3′-triiodothyronine (T_3) differs from thyroxine (T_4) in that

(A) T_3 is less potent than T_4
(B) over 70% of T_3 is produced outside the thyroid gland from T_4
(C) has a half-life of 7 days, whereas T_4 has a half-life of 1 day
(D) most of the T_3 in the plasma is bound to protein
(E) the total T_3 concentration in the plasma is more than 60 times that for T_4

611. Propylthiouracil prevents the formation of thyroxine (T_4) and triiodothyronine (T_3) from tyrosine and I^-; it therefore

(A) causes a decreased production of TRH (thyrotropin releasing hormone)
(B) causes a decreased production of TSH (thyroid-stimulating hormone)
(C) tends to produce an increased metabolic rate
(D) tends to prevent goiter
(E) tends to produce edema in the adult

612. Triiodothyronine (T_3)

(A) decreases the synthesis of m-RNA and the activity of myocardial adenylate cyclase
(B) decreases the sensitivity of adipose tissue to the fat-mobilizing action of epinephrine
(C) has a negative chronotropic action on the heart
(D) reduces the metabolic rate
(E) increases the activity of Na^+/K^+ pumps

613. Which of the following symptoms is NOT characteristic of hypothyroidism in the adult?

(A) a below normal mouth temperature
(B) a feeling of drowsiness
(C) increased body hair
(D) a deposition of mucoprotein in the subcutaneous and extracellular spaces that causes edema
(E) a negative nitrogen balance

614. Which one of the following symptoms is NOT characteristic of hypothyroidism in the adult?

(A) dry skin
(B) muscle weakness
(C) exophthalmos
(D) a deterioration of the thermoregulatory responses to cold
(E) a decreased ability to sweat in response to warming

615. Which of the following is INCORRECT? Cretinism

(A) is caused by prolonged, untreated hypothyroidism in the infant
(B) is associated with premature sexual development
(C) is frequently associated with symptoms that are not totally reversible
(D) is associated with mental retardation, potbellies, and enlarged protruding tongues
(E) can be caused by an iodine-deficient diet

616. Untreated goiter is associated with

(A) hyperthyroidism
(B) hypothyroidism
(C) euthyroidism
(D) hyperthyroidism and hypothyroidism
(E) hyperthyroidism, hypothyroidism, and euthyroidism

617. Hyperthyroidism causes

(A) an increase in concentration of thyroid-binding globulin (TBG)
(B) a positive nitrogen balance
(C) a potentiation of the action of catecholamines
(D) a decrease in urinary excretion of Ca^{++}
(E) lethargy

618. Which of the following is INCORRECT? A patient who, during surgery, has accidentally had all her parathyroid tissue removed will probably

(A) develop increased plasma free Ca^{++}
(B) develop skeletal muscle spasms
(C) show improvement in response to vitamin D
(D) show improvement in response to calcium gluconate injections
(E) develop personality changes and anxiety attacks

619. Parathormone (PTH)

(A) is a steroid
(B) secretion is decreased in response to a decreased level of plasma Ca^{++}
(C) inhibits the loss of phosphate by bone
(D) inhibits the loss of Ca^{++} from the bone
(E) increases the formation of 1,25-dihydroxycholecalciferol (vitamin D_3)

620. Parathormone

(A) by increasing the renal clearance of phosphate while decreasing the renal clearance of calcium, tends to produce a hypophosphatemia associated with hypercalcemia
(B) by increasing the renal clearance of calcium while decreasing the renal clearance of phosphate, tends to produce hypocalcemia associated with hyperphosphatemia
(C) produces hypercalcemia associated with hyperphosphatemia
(D) produces hypocalcemia associated with hypophosphatemia
(E) does not affect the renal clearance of phosphate and calcium

621. Increases in parathormone concentration result in all of the following EXCEPT

(A) increased absorption of Ca^{++} from the intestine
(B) increased excretion of Ca^{++} in the urine
(C) increased release of Ca^{++} into the blood by the osteocytes and osteoclasts
(D) increased Ca^{++} uptake by the bones
(E) increased blood Ca^{++}

622. Which of the following statements is INCORRECT? Calcitonin

(A) is released by the C cells of the thyroid
(B) is released in increased quantities in response to increases in plasma Ca^{++} and gastrin
(C) inhibits the release of calcium from the bone into the blood
(D) inhibits the action of parathormone on the bone
(E) decreases urinary excretion of Ca^{++}

DIRECTIONS (Questions 623 through 626): Each set of items in this section consists of a list of lettered options followed by several numbered words or phrases. For each numbered word or phrase, select the ONE lettered option that is most closely associated with it. Each lettered option may be selected once, more than once, or not at all.

Match the following hormones with their action on bone formation and mineralization.

(A) parathormone
(B) calcitonin
(C) growth hormone
(D) cortisol
(E) estrogen

623. This hormone inhibits the action of parathormone on bone and prevents osteoporosis

624. This hormone increases bone formation and mineralization by increasing collagen synthesis

625. This hormone decreases collagen synthesis of the bone and accelerates osteoporosis

626. Lack of this hormone is of little consequence

Answers and Explanations

610. (B) 70% of T_3 is produced by conversion from T_4. T_3 is also produced inside the thyroid gland from monoiodotyrosine (MIT). T_3 is about three times more potent than T_4. T_4 has a half-life of 7 days and T_3 one of 1 day. T_4 serves as a large reservoir of prohormone for T_3, as well as having some biologic activity of its own. There is more than 60 times more total T_4 in plasma than T_3. Eighty percent of it is bound to protein, whereas only 30% of T_3 is bound to protein.

611. (E) This goitrogen, by inhibiting the production of thyroid hormones (T_3 and T_4), increases the production of TSH (see Fig. 27–1).

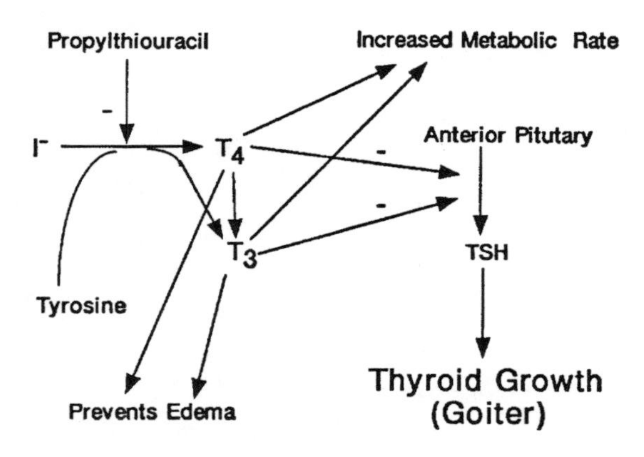

Figure 27–1. Actions of propylthiouracil on thyroid activity.

612. (E) T_3, by increasing adenylate cyclase activity, exerts a permissive action on the response of many cells (adipose tissue, heart, etc.) to epinephrine. It increases metabolic rate by increasing the activity of Na^+/K^+ pumps.

613. (C) The hair becomes coarse in about 76% of the cases, and there is a loss of hair in about 57% of the cases of hypothyroidism. In hypothyroidism, both the synthesis and the degradation of protein is reduced. The net result is that more nitrogen is lost from the body than is taken in. On the other hand, there are accumulations of extracellular proteins in a number of areas, which cause increases in colloid osmotic pressure in those areas. The mechanisms responsible for these changes are poorly understood.

614. (C) Exophthalmos is associated with hyperthyroidism of Graves' disease.

615. (B) Untreated hypothyroidism that occurs in the child, as opposed to the infant, produces a less severe series of changes and is called juvenile hypothyroidism. Cretinism is usually associated with amenorrhea and sterility. The degree of success in treating cretinism with thyroid hormones is inversely related to its duration. If permanent mental retardation is to be prevented, treatment must be begun before the symptoms become marked. Dietary iodine is essential for the production of T_3 and T_4.

616. (E) Goiter is an enlargement of the thyroid. It may occur in response to (1) a prolonged elevation in the TSH concentration of the plasma or (2) the production of LATS, an immunoglobulin. Some common causes of elevations in TSH are an iodine-deficient diet or the presence of goitrogens in the diet. In both cases, the goiter is associated with either hypothyroidism or euthyroidism. In euthyroid goiter due to iodine deficiency, the enlargement of the thyroid makes it possible for increased quantities of iodine to be removed from the blood, and thus it is a useful response to a stress. The production of LATS, unlike the production of TSH, is not decreased in response to thyroid hormones. Therefore, LATS produces a hyperthyroid goiter.

617. (C) Thyroid hormones potentiate the action of catecholamines. In hyperthyroidism, more T_3 and T_4 are protein bound but the concentration of TBG does not change. In both hypothyroidism and hyperthyroidism, there is a negative nitrogen balance. The catabolism of bone protein may in part contribute to the demineralization of bone and the resultant increase in urinary Ca^{++} seen in hyperthyroidism. Hyperthyroidism causes nervousness and hyperreflexia. There is also a muscle weakness and tendency to fatigue easily that might be confused with lethargy.

618. (A) Hypoparathyroidism causes a reduction of plasma Ca^{++} and as a result a hyperreflexia that causes an increase in muscle tone. Vitamin D will cause an increase in absorption of calcium from the GI tract and a decrease in calcium in the feces.

619. (E) PTH is a linear polypeptide with a molecular weight of 9500. An increase in plasma Ca^{++} causes a decrease in secretion of PTH. Calcium appears to be the primary determinant for the secretion of PTH. PTH facilitates the loss of calcium and phosphate by bone, and stimulates 1α-hydroxylate activity in the kidneys resulting in increased vitamin D_3 formation (see Fig. 27–2).

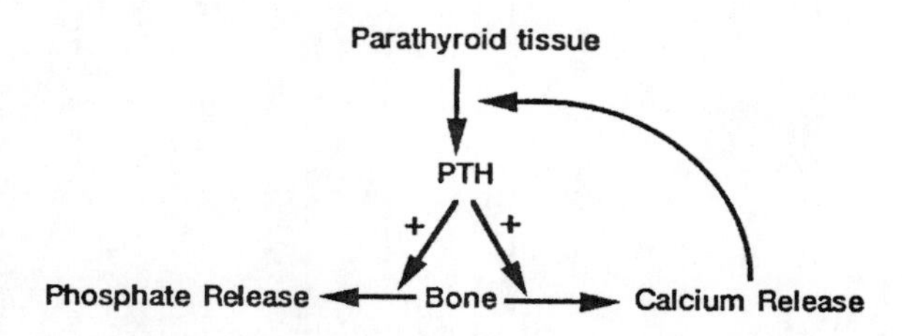

Figure 27–2. Actions of parathyroid hormone on bone.

620. (A) Parathormone increases Ca^{2+} and decreases phosphate blood levels (see Fig. 27–3).

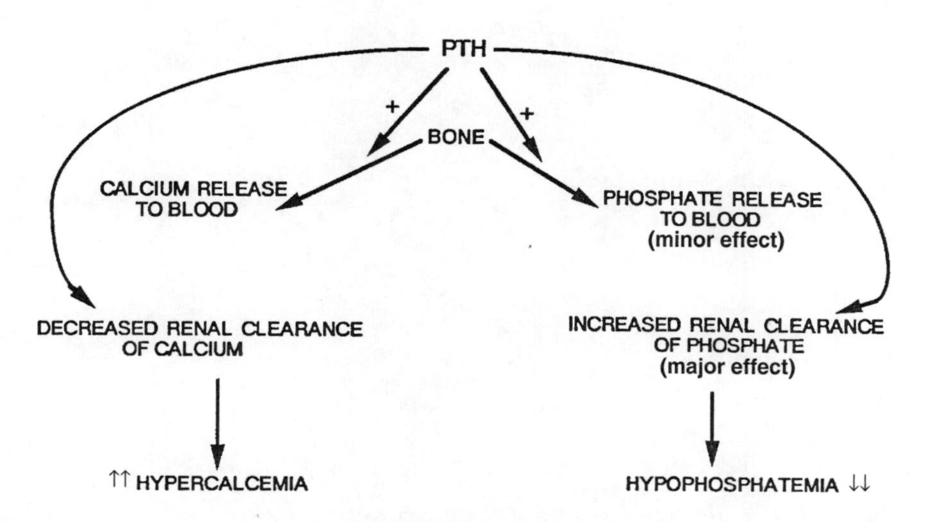

Figure 27–3. Actions of parathyroid hormone on blood Ca^{++} and phosphate.

621. (D) PTH increases intestinal Ca^{++} absorption by facilitating the synthesis of 1,25-dihydroxy-cholecalciferol (a metabolite of vitamin D) by the kidney. Frequently, the response to parathormone is a decrease in renal clearance of Ca^{++} and an increase in plasma Ca^{++} concentration, which collectively cause an increase in Ca^{++} excretion. The osteocytes seem to be responsible for the early rise and the osteoclasts the maintained rise in plasma Ca^{++}. Ca^{++} uptake by the bones is reduced.

622. (E) The role of calcitonin in human beings is not settled. It may play a role in skeletal development in the young and it may protect a mother from demineralization of her bones during pregnancy and lactation. It has proven useful in the treatment of Paget's disease (osteitis deformans).

623–626. (623-E, 624-C, 625-D, 626-B) Declining estrogen levels in postmenopausal women are a major factor in accelerated osteoporosis. Cortisol and growth hormone affect bone mineralization indirectly through their action on collagen synthesis.

CHAPTER 28

Hypothalamus, Pituitary, and Other Endocrine Glands

Questions

DIRECTIONS (Questions 627 through 630): Each set of items in this section consists of a list of lettered options followed by several numbered words or phrases. For each numbered word or phrase, select the ONE lettered option that is most closely associated with it. Each lettered option may be selected once, more than once, or not at all.

Match each of the hormones with their metabolic effects.

	Blood Sugar	Lipolysis	Nitrogen Balance
(A)	increased	increased	positive
(B)	increased	increased	negative
(C)	increased	decreased	positive
(D)	increased	decreased	negative
(E)	decreased	increased	positive
(F)	decreased	increased	negative
(G)	decreased	decreased	positive
(H)	decreased	decreased	negative

627. Cortisol

628. Growth hormone

629. Insulin

630. Glucagon

DIRECTIONS (Questions 631 through 637): Each of the numbered items or incomplete statements in this section is followed by answers or by completions of the statement. Select the ONE lettered answer or completion that is BEST in each case.

631. If the pituitary gland were removed and implanted on the kidney, the plasma concentration of

(A) adrenocorticotropic hormone would increase
(B) follicle-stimulating hormone would increase
(C) growth hormone would increase
(D) prolactin would increase
(E) thyrotropic hormone would increase

632. In human beings, total hypophysectomy does NOT cause

(A) a decreased secretion of cortisol
(B) a decreased secretion of testosterone
(C) cessation of the menstrual cycle
(D) a decrease in the hypoglycemic action of insulin
(E) infertility in the male

633. Total adrenalectomy is fatal in human beings, but hypophysectomy is not. This is because

(A) ADH from the pituitary and aldosterone from adrenals are antagonistic (ie, hypophysectomy causes a decreased production of both ADH and aldosterone
(B) GH and cortisol are antagonistic
(C) aldosterone secretion is not markedly reduced after hypophysectomy
(D) cortisol secretion is not markedly reduced after hypophysectomy
(E) the adrenal cortex hypertrophies after hypophysectomy

The adrenal gland can be divided into four functional units: (1) the capsule, (2) the zona glomerulosa, (3) the zona fasciculata and reticularis, and (4) the medulla.

634. The hypothalamus plays a major role in the control of

(A) only the zona glomerulosa
(B) only the zona fasciculata and reticularis
(C) only the medulla
(D) the zona glomerulosa and medulla
(E) the zona fasciculata and reticularis and the medulla

635. The pituitary gland plays a major role in the control of adrenal secretion by

(A) only the zona glomerulosa
(B) only the zona fasciculata and reticularis
(C) only the medulla
(D) the zona glomerulosa and medulla
(E) the zona fasciculata and reticularis and the medulla

636. Under resting conditions, the osmotic pressure of the plasma is 290 mOsm. If it decreases to 280 mOsm, all of the following statements are true EXCEPT

(A) there will be a reduced stimulation of the supraoptic neurons from the hypothalamus to the posterior pituitary
(B) ADH secretion will decrease
(C) systemic peripheral resistance to blood flow will decrease
(D) urine volume will increase
(E) the water permeability of the renal collecting ducts will decrease

637. Which of the following statements is INCORRECT? The *kidney* releases hormones or hormone-like agents

(A) in response to decreases in renal arterial pressure or constriction of renal artery
(B) in response to decreases in arterial O_2 tension
(C) in response to arterial aldosterone levels
(D) that indirectly facilitate the release of aldosterone
(E) that facilitate the production of erythrocytes

Answers and Explanations

627. **(B)** Cortisol increases blood sugar levels by promoting hepatic glycogenolysis. It enhances protein catabolism (negative N balance) (see Fig. 28–1).

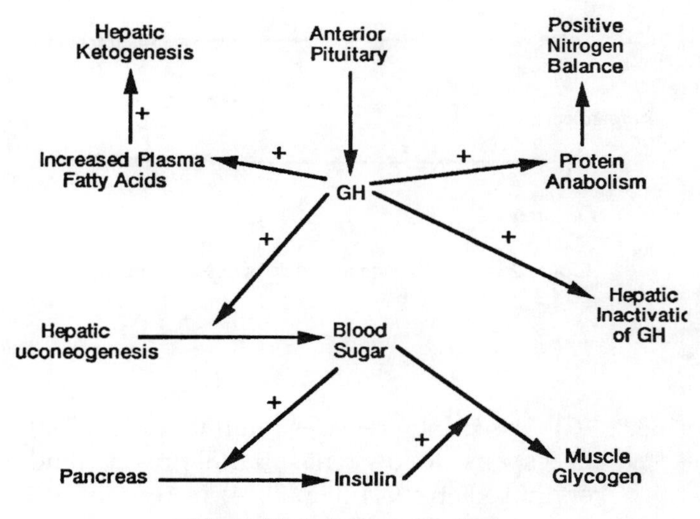

Figure 28–1. Actions of cortisol.

628. **(A)** GH is the only substance listed that increases blood sugar and causes a positive N balance. While preventing the catabolism of protein and amino acids, GH increases the availability of glucose, lipids, and ketone bodies in the blood (Fig. 28–2).

629. **(G)** Insulin is the only substance listed that decreases blood sugar. It increases the movement of blood sugar into most cells and in so doing increases carbohydrate catabolism and the formation of glycogen. Associated with this increased dependence on carbohydrates as a source of calories, there is a decreased concentration of ketone bodies and free fatty acids in the blood and a decreased protein catabolism.

630. **(B)** Glucagon and cortisol both increase blood sugar, cause a negative N balance, and are secreted in increased quantities during stress. However, glucagon is not essential for surviving a severe stress.

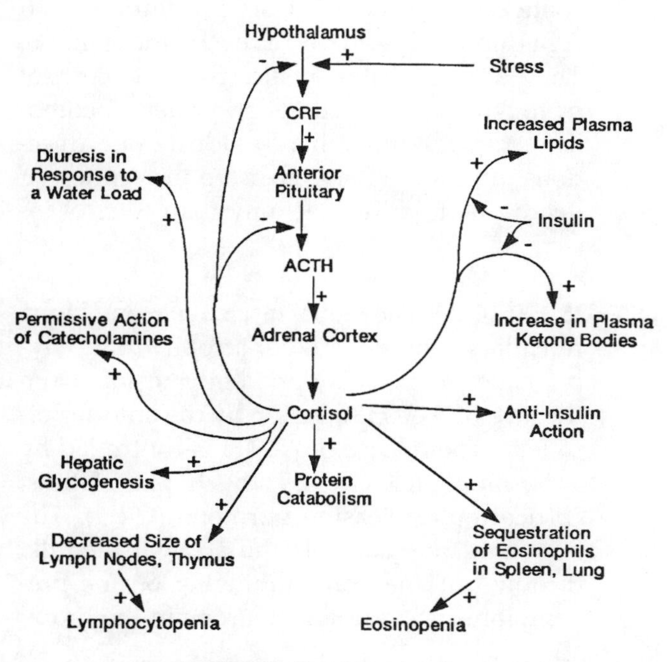

Figure 28–2. Actions of growth hormone.

631. **(D)** The hypothalamus produces releasing hormones that control the secretion of hormones from the anterior pituitary. The net effect of each of the releasing hormones, except those controlling prolactin secretion is to increase secretion. Therefore, when the anterior pituitary is removed from the influence of the hypothalamus and its PIH (prolactin inhibitory hormone), it releases more prolactin and less of its other secretions. What has been called PIH is the dopamine secreted by the tuberoinfundibular neurons into the portal hypophyseal vessels. Hormones that increase prolactin secretion include PRH (prolactin-releasing hormone), TRH (thyroid-stimulating hormone-releasing hormone), and estrogens.

632. **(D)** Hypophysectomy causes exaggerated sensitivity to insulin due to loss of pituitary ACTH and growth hormone. The most important action of pituitary ACTH is to stimulate the secretion of cortisol and other glucocorticoids. Pituitary LH stimulates the secretion of testosterone. Pituitary FSH and LH regulate the menstrual cycle. Pituitary FSH stimulates spermatogenesis.

633. **(D)** Hypophysectomy causes a marked decrease in the release of cortisol but little or no change in the secretion of aldosterone. Both ADH and aldosterone reduce urine volume. They are not antagonistic. GH and cortisol have antagonistic actions on protein metabolism. On the other hand, aldosterone injections are more effective in the prevention of death due to adrenalectomy than cortisol injections.

634–635. **(634-A)** The zona fasciculata and zona reticularis show marked atrophy after hypophysectomy. This apparently results from the lack of ACTH secretion by the anterior pituitary. The release of ACTH is controlled by a hormone released by the hypothalamus, corticotropin-releasing hormone (CRH). The hypothalamus controls the adrenal medulla through interneurons impinging on the preganglionic sympathetic neurons in the thoracolumbar cord. Hypophysectomy produces little or no change in the adrenal medulla. All the zones of the cortex release glucocorticoids and sex hormones, but the zona glomerulosa is the major source for aldosterone. It is the only zone in the cortex that does not atrophy within the first 6 weeks of hypophysectomy.

636. **(C)** ADH is also called vasopressin. In large doses, vasopressin produces vasoconstriction, but it is doubtful that a healthy person ever produces enough to cause changes in peripheral resistance. Its major action is to decrease water loss in the urine by increasing the water permeability of the collecting ducts.

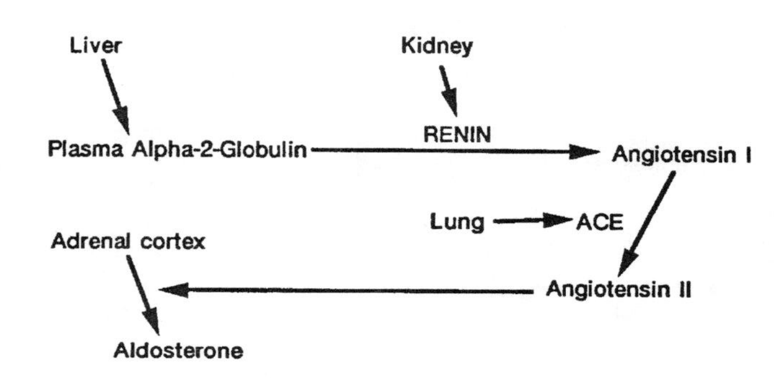

Figure 28–3. Formation and actions of angiotensin.

637. **(C)** The kidney releases renin (see Fig. 28–3) in response to low renal arterial pressure and erythropoietin (see Fig. 28–4) in response to low arterial O_2 tension.

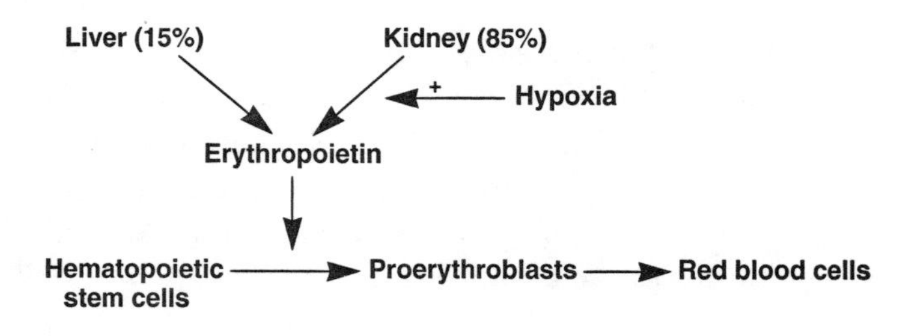

Figure 28–4. Formation and actions of erythropoietin.

CHAPTER 29

Fertilization, Growth, and Maturation

Questions

DIRECTIONS (Questions 638 through 653): Each of the numbered items or incomplete statements in this section is followed by answers or by completions of the statement. Select the ONE lettered answer or completion that is BEST in each case.

638. Which of the following statements is INCORRECT? Human spermatozoa normally

(A) are not motile in the epididymis but are activated by the secretions of the prostate and other accessory glands of the male
(B) are produced at an accelerated rate when testicular temperature is raised from 35 to 37°C
(C) contain 23 chromosomes
(D) contain either an X or a Y chromosome but not both
(E) can survive in the female reproductive tract for 1 or 2 days or longer

639. Which of the following statements is INCORRECT? Human spermatozoa normally

(A) at a concentration of 1 million/mL of ejaculate will fertilize the ovum
(B) contain enzymes in their head that facilitate the penetration of the ovum
(C) once one has penetrated an ovum, initiate a process that prevents other spermatozoa from penetrating that ovum
(D) are released in greater numbers in the ejaculate of a 21-year-old after 3 days of abstinence than after 1 day of abstinence
(E) take about 74 days to change from a relatively undifferentiated spermatogonium into a mature spermatozoan

640. Which of the following statements is INCORRECT? Human spermatozoa normally

(A) remain motile in the female reproductive tract for a period of less than 15 minutes after ejaculation during intercourse
(B) leave the penis, suspended in a liquid (semen), most of which comes from the prostate gland and seminal vesicles immediately prior to and during ejaculation
(C) derive from spermatogonia, cells in the seminiferous tubules that contain the diploid number of chromosomes
(D) are stored in the seminiferous tubules, rete testis, epididymis, and vas deferens
(E) may remain viable in the epididymis for periods up to 60 days

641. A child with a genotype of

(A) XX/XY will develop into a true hermaphrodite
(B) XY will contain a Barr body in its cells
(C) XXX will develop excessively wide hips
(D) XO will develop into a normal female
(E) XXY will develop into a true hermaphrodite

642. Which of the following statements is INCORRECT? A premature infant is more likely than a full-term infant to

(A) suffer from severe jaundice of hepatic origin
(B) lack fontanels
(C) suffer from anemia
(D) lack surfactant
(E) have difficulty regulating body temperature

643. In the newborn

(A) the ability to manufacture antibodies is poorly developed and will not be comparable to that of the adult until approximately the age of 12 months
(B) there is practically no resistance to infection
(C) lymphocyte precursors have not yet been changed to T lymphocytes by the thymus
(D) the brain cells are less tolerant of a lack of O_2 than in the adult
(E) the mechanism for concentrating urine is more efficient than in the adult

644. Acromegaly is a condition caused by

(A) an excessive release of growth hormone, which causes, in the adult, enlarged hands and feet and protrusion of the lower jaw
(B) an excessive release of growth hormone, which causes, in the adult, an elongation of the humerus, femur, tibia, and fibula
(C) an excessive release of thyroid hormone, which causes, in the adult, enlarged hands and feet and a protrusion of the lower jaw
(D) an excessive release of thyroid hormone, which causes, in the adult, an elongation of the humerus, femur, tibia, and fibula
(E) an inadequate release of thyroid hormone

645. Two years prior to the occurrence of puberty

(A) the gonads are insensitive to pituitary gonadotropins
(B) the pituitary and hypothalamus are devoid of gonadotropins
(C) the concentration of adrenal androgens in the blood is markedly lower than in boys and girls at 15 years of age
(D) there is little pulsatile release of GnRH from the hypothalamus
(E) the pituitary gland does not respond to pulsatile release of GnRH.

646. Puberty before the age of 10 years

(A) may be caused by destruction of the pituitary
(B) causes a delay in the closure of the epiphyseal plates
(C) in over 80% of the cases pathologic tumors are found in the patient
(D) may be caused by hypothalamic lesions
(E) is extremely rare

647. During puberty the normal female experiences

(A) increased FSH, LH, and estrogen
(B) decreased FSH and LH, and increased estrogen
(C) increased FSH and LH, and decreased estrogen
(D) decreased FSH, LH, and estrogen
(E) no change in estrogen

648. Which of the following statements is INCORRECT? Erection of the penis

(A) occurs in response to a reflex stimulation of parasympathetic neurons in the sacral spinal cord
(B) is initiated by afferent neurons from the glans penis or descending fibers in the cord
(C) may occur 5 to 10 years prior to puberty
(D) is a result of arteriolar dilation in the penis
(E) requires intact brain stem function

649. Six months after castration, an adult man will have

(A) a decreased secretion of LH
(B) atrophy of the prostate gland

(C) abolition of the libido
(D) an increased pitch of the voice
(E) breast enlargement

650. Which of the following statements is INCORRECT? Follicle-stimulating hormone

(A) in the presence of LH, facilitates the release of estrogen by the theca interna of the Graafian follicle
(B) facilitates spermatogenesis
(C) increases tenfold during pregnancy
(D) release is modulated by impulses impinging on the hypothalamus
(E) is secreted by the anterior pituitary

651. Which of the following statements is INCORRECT?

(A) castration of the adult man causes a prompt increase in the concentration of circulating FSH and LH
(B) testosterone decreases the release of LH and FSH
(C) testosterone is secreted by Leydig cells
(D) inhibin decreases the release of FSH
(E) inhibin is secreted by Leydig cells

652. In normal adult human subjects, prolactin

(A) in the presence of ovarian and adrenal hormones, promotes the growth and development of the mammary glands
(B) facilitates the secretory activity of the corpus luteum
(C) facilitates the development of the Leydig cells of the testis
(D) facilitates the release of FSH
(E) is suppressed during pregnancy

653. Which of the following statements is INCORRECT?

(A) androgens are in greater concentration in the blood of the male fetus than in the blood of the male child
(B) androgens are formed by the seminiferous tubules of the testis
(C) androgens are secreted in small quantities in the adult female
(D) androgens in men decrease in concentration after the age of 30
(E) LH has little effect on the secretion of androgens by the adrenal cortex

DIRECTIONS (Questions 654 through 656): Each set of items in this section consists of a list of lettered options followed by several numbered words or phrases. For each numbered word or phrase, select the ONE lettered option that is most closely associated with it. Each lettered option may be selected once, more than once, or not at all.

(A) 17-OH progesterone
(B) chorionic somatomammotropin
(C) syncytiotrophoblast cells
(D) decidual cells
(E) pituitary prolactin
(F) progesterone
(G) estriol (16-OH estradiol)
(H) estradiol
(I) DHEA sulfate
(J) zona reticularis of the adrenal cortex
(K) prostaglandins
(L) oxytocin
(M) chorionic gonadotropin
(N) cytotrophoblast
(O) decidual prolactin

654. Labor is probably induced by pituitary production of this substance in response to severe stretch of the uterine cervix

655. Within 5 to 7 days after ovulation and conception, the implanting embryo begins producing this substance, which is the hormone that saves the corpus luteum from regression

656. The secretory function of the corpus luteum can be monitored by determining the levels of this in the plasma or serum

Answers and Explanations

638. **(B)** The descent of the testes into the scrotal sac sends them into an environment about 2°C below the temperature in the abdomen. If they remain in the abdominal cavity or if their temperature in the scrotal sac is changed from 35 to 37°C, they will stop producing spermatozoa, and if this situation persists, degeneration of the tubular walls and sterility will eventually occur. Survivals of sperm in the female reproductive tract of up to 6 days have been reported.

639. **(A)** There are normally 100 million sperm per mL of ejaculate. About half the men with counts between 20 and 40 million per mL are sterile. Counts below 20 million per mL are almost always sterile.

640. **(A)** They remain motile for periods in excess of 2 hours. It probably takes a spermatozoan over an hour to migrate up to the ovum. One cause of sterility in the male is spermatozoa with short-lived motility. About 80% of the seminal fluid comes from the seminal vesicles and prostate.

641. **(A)** The true hermaphrodite has some cells that are female (XX) and some that are male (XY). It may contain both an ovary and a testis. Females (XX) contain Barr bodies (usually seen as an extra quantity of chromatin attached to the nuclear membrane in the karyoplasm) in each of their nucleated cells (the cells of mucous membranes, for example). Females with XXX genotype will develop normally. In XO genotype (Turner's syndrome), the individual has the external genitalia of an immature female and gonads that are either rudimentary or absent (incidence = 0.03%). A male with XXY genotype will develop normal external genitalia. He will have abnormal seminiferous tubules and, since he has two X chromosomes, will be chromatin-positive (ie, have a Barr body). This condition (Klinefelter's syndrome) has an incidence of 0.26% and is frequently associated with mental retardation.

642. **(B)** The bone that forms the cranium is membranous bone. It starts as a soft connective tissue membrane that becomes infiltrated with osteoblasts. In both the premature and mature newborn infant, calcification of this membrane will be incomplete. In other words, both will have soft spots called fontanels in the cranium. The fontanels normally have disappeared in the 18-month-old infant. The newborn will frequently be jaundiced. This is probably because the glucuronide conjugating system of the liver has not fully developed. The jaundice will usually disappear by the 10th day of life. In the premature infant the problem is much more severe. His deficiency in hepatic glucuronyl transferase activity may lead to a sufficiently serious jaundice to produce central nervous system damage (kernicterus). The normal infant is born with a reserve of iron that will last him about 6 months. This is important because cow's milk is deficient in iron. The premature infant, on the other hand, is frequently deficient in iron because these stores are laid down late in pregnancy. Lack of surfactant can lead to lung collapse (respiratory distress of the newborn). Temperature regulating mechanisms mature in late pregnancy and early infancy.

643. **(A)** The newborn has a good titer of light gamma globulins (IgG), which he has received from the maternal circulation. These will have disappeared by 10 weeks of age, but by this time he is producing his own. At birth, however, he does lack the heavier gamma globulins (IgM and IgA). Thus, he is susceptible to infection by gram-negative organisms but resistant to many other types of infection. B lymphocytes produce antibodies. T lymphocytes promote cellular immunity. Most of the T lymphocytes are produced by the thymus prior to birth and during the several months after birth. If the thymus is removed several months before birth, the body loses most of its ability to reject transplanted organs. The kidney of the newborn has one half the concentrating ability of the kidney in the adult.

644. **(A)** The hypersecretion of GH that produces acromegaly occurs after most of the long bones have ceased to grow. It will not, therefore, cause increases in the length of the femur and humerus. It may increase the length of some bones that have not ceased to grow yet, but it will in all cases increase the size of soft tissue (the tongue and abdominal viscera, for example) and will increase the thickness of a number of bones. This latter effect is particularly obvious in the hands, feet (the subject may have to wear a shoe that is size 14 or larger), the nose (it may be twice normal size), and the mandible.

645. **(D)** Prior to puberty the gonads are sensitive to gonadotropins and there are gonadotropins in the pituitary and hypothalamus, but they are not being released. Pulsatile release of GnRH by the hypothalamus initiates puberty.

646. **(D)** Removal of the pituitary will prevent the onset of puberty. A lesion in the hypothalamus, on the other hand, may cause a premature puberty. There would be a premature growth spurt followed by a premature closure of the epiphyses. The individual who experiences premature puberty is usually shorter than the average adult. It has been estimated that puberty prior to 10 years of age is the result of a pathologic condition in less than 25% of the cases. There is a wide range of normal for the onset of puberty. Menarche frequently occurs between the ages of 9 and 17.

647. **(A)** There is an increased secretion of estrogen, follicle-stimulating hormone (FSH), and luteinizing hormone (LH) during puberty.

648. **(E)** Erection has been noted at all ages ranging from 1 day old to 80 years old. It occurs as reflex stimulation of sacral autonomic neurons and does not require intact brain stem function. However, damage to the brain stem might impair psychogenic erections.

649. **(B)** Testosterone inhibits the release of luteinizing hormone-releasing hormone (LRH). Libido may decrease, but it is not abolished. The growth of the larynx that occurs during puberty, unlike that of the prostate, is permanent.

650. **(C)** FSH plays an important role in both men and women in facilitating the production of mature eggs (ie, sperm and ova). The hypothalamus secretes a releasing hormone that facilitates FSH secretion. Its release is affected by estrogens, stress, and possibly other influences. FSH is secreted in greatly reduced quantities during pregnancy.

651. **(E)** Inhibin is a glycoprotein secreted by Sertoli cells and has a strong inhibitory effect on FSH release by the anterior pituitary.

652. **(A)** Prolactin is also one of the important hormones responsible for lactation by the mother after the birth of her child. Its serum level increases tenfold during pregnancy. Prolactin has no well-defined function in the human male.

653. **(B)** The Leydig cells (interstitial cells adjacent to the seminiferous tubules) secrete androgens. Testosterone is, at least in part, responsible for the growth of the penis and scrotum during prenatal development. The adult female produces between 0.27 and 0.35 mg of testosterone per day. Some is from the ovary, some from the adrenal cortex, and some from the conversion of androstenedione by other tissues. These decreasing concentrations of androgens are gradual in men, unlike the decrease in concentration of estrogen and progesterone in women after menopause.

654. **(L)** Oxytocin is released by the posterior pituitary gland in response to stretch of the cervix and potentiates the weaker Braxton Hicks contraction.

655. **(M)** Chorionic gonadotropin prevents regression of the corpus luteum and is measured by immunoassays in pregnancy tests.

656. **(F)** The function of the corpus luteum is assayed by measuring plasma or serum progesterone. 17-hydroxyprogesterone is a metabolite of progesterone which accumulates in 21-hydroxylase deficiency (congenital adrenal hyperplasia).

CHAPTER 30

Female Reproductive System

Questions

DIRECTIONS (Questions 657 through 676): Each of the numbered items or incomplete statements in this section is followed by answers or by completions of the statement. Select the ONE lettered answer or completion that is BEST in each case.

657. In the normal, healthy 25-year-old woman, menstruation

(A) occurs 1 to 2 days after ovulation
(B) always occurs every 4 weeks except during or immediately after pregnancy
(C) occurs several hours after the formation of a corpus luteum in the ovary
(D) is a period of secretion by the endometrium of the uterus
(E) occurs 1 to 2 days after the formation of a corpus albicans in the ovary

658. In the normal, healthy 25-year-old woman who is not pregnant, menstruation

(A) lasts about 1 day during each 28-day cycle
(B) occurs 1 or 2 days after an increase in the estrogen and progesterone levels of the blood has begun
(C) is associated with a blood loss of about 30 mL
(D) is associated with a dilation of the basal segment of the spiral artery of the endometrium
(E) is associated with an increased risk of uterine infection

659. Which of the following hormones exerts little or no control over the endometrium of the uterus in its proliferative phase, but during its secretory phase is directly responsible for the changes that occur?

(A) progesterone
(B) follicle-stimulating hormone
(C) estrogen
(D) prolactin
(E) luteinizing hormone

660. In a normal, healthy 25-year-old woman

(A) the plasma LH is at its lowest concentration during the 2 days prior to ovulation
(B) the plasma FSH is at its lowest concentration during the 2 days prior to ovulation
(C) the plasma estrogen is at its lowest concentration during the 2 days prior to ovulation
(D) ovulation is followed by a decline in plasma estradiol
(E) ovulation is followed by a decline in plasma progesterone

661. In a normal, healthy 25-year-old woman with a menstrual cycle of 28 days

(A) the proliferative phase of the uterus is caused by estrogen produced by the Graafian follicle
(B) menstruation is caused by progesterone from the corpus luteum
(C) oral estrogen and/or progesterone will cause an enlargement of the ovary and an increase in production of mature Graafian follicles
(D) the concentration of estradiol in the plasma begins to fall at ovulation and continues to decrease until menstruation
(E) the concentration of progesterone in the plasma begins to fall at ovulation and continues to decrease until menstruation

662. In a normal 25-year-old woman, 4 days prior to the onset of menstruation there are

(A) decreasing plasma concentrations of estrogen and progesterone because of a decreased release of prolactin by the pituitary
(B) decreasing plasma concentrations of estrogen and progesterone because the corpus luteum, in the absence of pregnancy, has a short life span
(C) decreasing plasma concentrations of estrogen and progesterone because of decreasing concentrations of LH
(D) increasing concentrations of progesterone
(E) increasing concentrations of estrogen and progesterone

663. Which of the following statements is INCORRECT? Combined oral contraceptives

(A) will prevent menstruation
(B) depress the release of gonadotropins by the anterior pituitary
(C) have a failure rate of less than 0.1%
(D) cause an increased extracellular fluid volume
(E) increase the risk of thrombosis

664. Which of the following statements is INCORRECT?

(A) ovaries are essential for cyclic uterine activity
(B) ovaries start to form ova at puberty
(C) the cervical mucus becomes less viscous at the time of ovulation
(D) a foreign body in the uterus prevents implantation
(E) in most women, 14 days after the onset of menstruation is a more fertile period than 22 days after the onset of menstruation

665. When fertilization occurs, there is a missed menstrual period. This first missed period results from the fact that after fertilization

(A) the corpus luteum degenerates
(B) a trophoblast is formed which secretes estrogen and progesterone
(C) a trophoblast is formed which secretes gonadotropins
(D) the ovaries decrease their production of estrogen and progesterone
(E) negative feedback of estrogen and progesterone on pituitary FSH and LH release

666. During most of pregnancy, the maintenance of human gestation is dependent upon the secretions of the

(A) corpus luteum of the ovary
(B) anterior pituitary
(C) corpus luteum and anterior pituitary
(D) placenta
(E) adrenal cortex

667. During pregnancy, human chorionic gonadotropin (hCG)

(A) slowly increases in concentration throughout pregnancy
(B) starts to decrease in concentration at about the middle of the first trimester of pregnancy
(C) causes a decrease in concentration of estrogen and progesterone in the blood

(D) increases in concentration throughout pregnancy and, while increasing, decreases the concentration of estrogen and progesterone
(E) increases in concentration throughout pregnancy and, while increasing, increases the concentration of estrogen and progesterone

668. All of the following hormones are produced by the placenta EXCEPT

(A) human chorionic gonadotropin
(B) oxytocin
(C) human chorionic somatomammotropin
(D) estrogen
(E) progesterone

669. During pregnancy,

(A) spontaneous contractions of the uterus are usually absent until the 36th week
(B) the breasts enlarge due to the release of prolactin from the anterior pituitary
(C) the O_2 tension in the umbilical artery exceeds that in the umbilical vein
(D) the blood in the umbilical vein has an O_2 tension similar to or greater than the blood in an average systemic vein of the mother
(E) hCG reaches a peak after 2 to 3 months gestation, then declines

670. From the 10th week of pregnancy until delivery there is, in the mother, a progressive increase in the concentration of all of the following hormones EXCEPT

(A) plasma estriol
(B) plasma human chorionic somatomammotropin (hCS)
(C) urinary pregnanediol
(D) plasma chorionic gonadotropin (hCG)
(E) progesterone

671. The events that initiate parturition (labor) are not well defined for human beings. We do know, however, that during the last day of pregnancy

(A) oxytocin is released by the posterior pituitary and decreases the frequency of uterine contractions
(B) stretching the uterine cervix is a potent stimulus for oxytocin release
(C) prostaglandin formation in the decidua of the uterus is depressed
(D) prostaglandin decreases uterine contractions
(E) progesterone facilitates uterine contractions

672. Which of the following statements is correct?

(A) during the first stages of labor, uterine contractions occur every 15 minutes and last about 10 to 15 seconds
(B) as labor progresses, the contractions occur less frequently but are more forceful
(C) the chorion and amnion generally rupture 1 to 2 days prior to labor and, as a result, the amnionic fluid rushes out the vagina
(D) in an ideal pregnancy, an average woman should gain less than 15 pounds
(E) sulfonamides are a good choice of antibiotics for pregnant women

673. Which of the following statements is correct?

(A) in a lactating mother, within 24 hours after the end of parturition the uterus stabilizes within 10% of its weight prior to pregnancy
(B) it is usually "safe" to have coitus during menstruation, the first 8 months of pregnancy, and 1 month after parturition
(C) in the postpartum period, most women who breast feed their children have less interest in coitus than those who do not nurse
(D) there is a reduced interest in sexual intercourse in the woman during the second trimester of pregnancy
(E) sympathetic stimulation enhances oxytocin secretion and lactation

674. Which of the following statements is INCORRECT? In a 24-year-old woman

(A) the breasts enlarge during pregnancy because of the high circulating levels of estrogen and progesterone
(B) the breasts enlarge during pregnancy because of the synergistic actions of estrogen and prolactin
(C) after parturition an abrupt decrease in estrogen concentration initiates lactation
(D) after parturition the progesterone concentration increases abruptly
(E) suckling facilitates the release of oxytocin from the posterior pituitary

675. Which of the following statements is INCORRECT? In the lactating woman

(A) oxytocin causes the contraction of the myoepithelial cells
(B) prolactin causes the secretion of milk
(C) nursing inhibits the secretion of FSH and LH
(D) oxytocin is essential for milk ejection
(E) menstruation is more heavy than in women who do not nurse

676. The postmenopausal woman has

(A) an elevation of gonadotropins caused by a decreased secretion of ovarian steroids
(B) a tendency toward hot flashes as a result of increased secretion of estrogen
(C) elevations in plasma concentrations of gonadotropins and ovarian steroids
(D) an incapacity to achieve orgasm
(E) a decreased concentration of gonadotropins in the plasma

DIRECTIONS (Questions 677 through 682): Each set of items in this section consists of a list of lettered options followed by several numbered words or phrases. For each numbered word or phrase, select the ONE lettered option that is most closely associated with it. Each lettered option may be selected once, more than once, or not at all.

Questions 677 through 679

The following diagram (Fig. 30–1) explains the feedback regulation of sex hormones in a healthy 25-year-old woman. Match each of the hormones with the following list.

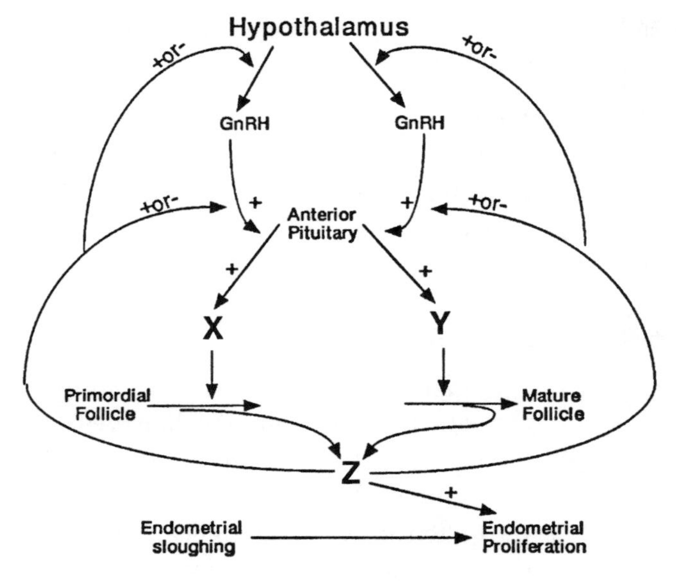

Figure 30–1. Regulation of sex hormones.

(A) estrogen
(B) testosterone
(C) progesterone
(D) hCG
(E) FSH
(F) LH

677. Hormone X

678. Hormone Y

679. Hormone Z

Questions 680 through 682

The following diagram (Fig. 30–2) illustrates some of the relationships that exist in the lactating woman who is nursing a child. Match it to the following list of hormones.

(A) estrogen
(B) progesterone
(C) FSH
(D) LH
(E) oxytocin
(F) prolactin
(G) dopamine

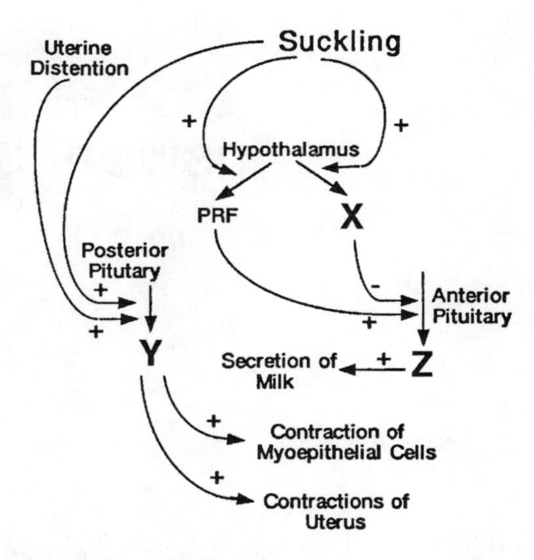

Figure 30–2. Physiological relationships in the lactating woman (PRF: prolactin releasing factor).

680. Hormone X

681. Hormone Y

682. Hormone Z

Answers and Explanations

657. (E) In the absence of fertilization, the corpus luteum degenerates into a structure, the corpus albicans. Since this structure, unlike the corpus luteum, does not secrete estrogen or progesterone, menstruation will ensue. In a 28-day cycle, menstruation begins about 14 days after ovulation. Various stressful situations (death in the family, worry, an automobile accident) may cause a missed period. The corpus luteum, through the estrogen and progesterone it produces, prevents menstruation. The secretory phase of the cycle begins on about the 14th day. Menstruation involves the sloughing of about two-thirds of the endometrium.

658. (C) The total volume of menstrual fluid lost per month is about 70 mL. Of this, about 30 mL is liquid blood. Menstrual bleeding is generally greatest during the first 3 days and ends about the 5th day. When the corpus luteum stops releasing these hormones, there is a decreased blood flow to the endometrium, due to constriction of the spiral arteries, resulting in necrosis and endometrial sloughing. Tremendous numbers of leukocytes are released along with the necrotic material, rendering the uterus highly resistant to infection during menstruation.

659. (A) Progesterone is released in increased quantities during the second half of the menstrual cycle after the mature follicle has ruptured and part of it has changed into a corpus luteum. FSH exerts its influence on the uterus by controlling the release of hormones from the ovary (follicle and corpus luteum). Estrogen controls the proliferative phase of the endometrium. Prolactin probably exerts no influence on the endometrium. LH, like FSH, exerts its influence on the endometrium by controlling the ovary.

660. (D) The estrogen concentration falls temporarily after ovulation as the remains of the ruptured follicle reorganize into a corpus luteum, which will also secrete estrogens. The high LH concentration is responsible for initiating the ovulation. The FSH concentration increases prior to ovulation. The progesterone concentration will rise with the formation of the corpus luteum.

661. (A) Estrogen promotes proliferation of the endometrium. Menstruation occurs when the corpus luteum stops producing progesterone and estrogen. Estrogen acts on the pituitary and/or hypothalamus to decrease the production of FSH. Progesterone, probably through a permissive action, facilitates this inhibition. In other words, these two hormones, by inhibiting FSH release, tend to cause ovarian atrophy and prevent the development of a mature Graafian follicle and ovulation. This is how they act in birth control pills. After ovulation, the corpus luteum is formed. It produces enough estrogen to cause a second rise in plasma estrogen.

662. (B) The best explanation for the conversion of the corpus luteum into a corpus albicans after only 12 days of function is its own intrinsic characteristics. LH is capable of prolonging the life of the corpus luteum and facilitating its production of progesterone, but in the absence of pregnancy sufficiently large concentrations of LH do not appear.

663. (C) Combined oral contraceptives contain estrogen and a progesterone. The usual way to administer these pills is to either give a few blank pills each month or to stop taking these pills for a few days each month. It is the withdrawal of estrogen and progesterone that causes menstruation. The typical failure rate is 3%. Estrogens in large concentrations probably act on the kidney to increase Na and water retention. They also significantly increase the risk of venous thrombosis.

664. (B) Apparently no additional ova are formed after birth. At puberty, a ripening process begins. This is a situation different from what we find in the male. The use of intrauterine devices is a more effective means of contraception than condoms. It is less effective than the use of oral contraceptives. Ovulation occurs at approximately 14 days after the beginning of menstruation.

665. (C) The corpus luteum hypertrophies and releases increased quantities of estrogen and progesterone. The first missed menstrual period is due to the human chorionic gonadotropin (hCG) released by the developing placenta. hCG facilitates the continued growth of the corpus luteum and its release of increasing quantities of estrogen and progesterone. As the placenta develops, it will eventually produce enough steroids to make the corpus luteum unnecessary.

666. (D) Until the placenta has formed and matured, the corpus luteum helps to prevent the sloughing of the endometrium of the uterus. After the placenta has formed, it produces chorionic gonadotropins which control the production of estrogens and progesterone by the corpus luteum. After it has matured, it produces its own estrogens and progesterone and the corpus luteum undergoes degenerative changes. The anterior pituitary has its effect on pregnancy through its control over the secretions of the ovary. Although the adrenal cortex of the fetus, and possibly of the mother, secretes large quantities of hormones during pregnancy, there is no evidence that these play the key role that the hormones of the placenta do.

667. (B) hCG increases the production of estrogen and progesterone by the corpus luteum. The corpus luteum maintains pregnancy until the placenta starts producing sufficient quantities of estrogen and progesterone to maintain the pregnancy without the help of the corpus luteum.

668. (B) Oxytocin is produced by the posterior pituitary. Human chorionic gonadotropin (hCG) facilitates the release of estrogen and progesterone by the corpus luteum of the ovary. Human chorionic somatomammotropin (hCS) promotes mammary growth, a positive nitrogen balance, and has an action similar to growth hormone. Estrogen (estradiol) promotes growth and maintenance of the endometrium of the uterus. Progesterone is responsible for the progestational changes in the endometrium.

669. (E) Uterine contractions occur throughout pregnancy. During the last weeks of pregnancy they become more rhythmic, occurring as frequently as every 1 to 20 minutes. Estrogen and progesterone prevent the release of prolactin from the pituitary. The umbilical vein carries oxygenated blood away from the placenta and therefore has more O_2 than the umbilical artery. The blood in the umbilical vein apparently has a Po_2 well below 40 mm Hg.

670. (D) The placenta produces and releases progressively greater quantities of estrogen, hCS, and progesterone from the 8th week throughout the pregnancy. Pregnanediol is the principal metabolite of progesterone. hCG is at its highest concentration in the eighth week of pregnancy.

671. (B) Oxytocin, during labor, increases uterine contractions by (1) increasing the production of prostaglandins in the decidua and (2) a direct action on the uterus. Progesterone usually inhibits uterine contractions.

672. (A) As labor progresses, the contractions eventually occur every 2 to 4 minutes and last 45 to 60 seconds each. The membranes generally rupture when the head of the fetus is penetrating the cervix of the uterus (the first stage of labor) but may break days or weeks before the onset of labor or not break until after the child is born. On an average, the woman gains 21 pounds. This gain is considered desirable. Sulfonamides are teratogenic in the third trimester.

673. (B) After delivery, the uterus weighs about 1 kg. After 4 to 5 weeks, it has involuted and returned to a weight of 50 g. Women who breast feed their infants usually have more interest in sex. Also, there is an increased interest in sex during the first half of pregnancy. During the third trimester, there is usually a decreased interest in sex.

674. (D) The expulsion of the placenta during parturition eliminates the major source of estrogen and progesterone during pregnancy. Suckling inhibits the release of PIH (prolactin-inhibiting factor). In other words, suckling causes the release of prolactin by withdrawing the inhibition of PIH secretion.

675. (E) About half of all nursing mothers do not menstruate until their child is weaned. Mothers who do not nurse have their first menstrual period about 6 weeks after delivery.

676. (A) The decreased secretion of the ovarian steroid estrogen results in an elevation in the plasma concentration of gonadotropins and a tendency toward hot flashes.

677–679. (677-E, 678-F, 679-A) When estrogen concentration is high (in midcycle), there is a positive feedback relationship between estrogen and the gonadotropins (FSH and LH). During the rest of the cycle, when estrogen concentrations are lower there is a negative feedback. This explains the "+ or –" notation in the figure.

680–682. (680-G, 681-E, 682-F) Dopamine, also known as prolactin-inhibiting hormone (PIH) suppresses the secretion of prolactin from the anterior pituitary. Oxytocin is secreted by the posterior pituitary.

PART IX

Sensation and Integration

CHAPTER 31

Vision

Questions

DIRECTIONS (Questions 683 through 698): Each of the numbered items or incomplete statements in this section is followed by answers or by completions of the statement. Select the ONE lettered answer or completion that is BEST in each case.

683. A lens that brings parallel rays of light to focus 2 cm from its center has a focal length of 0.02 m and a refractive power of

(A) less than 1 diopter
(B) between 1 and 10 diopters
(C) between 11 and 25 diopters
(D) between 26 and 46 diopters
(E) greater than 47 diopters

684. Which of the following statements is INCORRECT? An individual with emmetropic eyes, in order to see an object distinctly

(A) does not have to accommodate if the object is 100 feet away
(B) does not have to accommodate if the object is 10 feet away
(C) must stimulate parasympathetic neurons to his ciliary muscle if the object is 5 feet away
(D) must increase the tension on the suspensory ligament of his eyes if the object is 5 feet away
(E) must increase the refractory power of his lens if the object is 5 feet away

685. Which of the following statements is INCORRECT? In accommodation to a near object

(A) the lens becomes more convex
(B) there is a convergence of the two eyes owing to the stimulation of somatic efferent neurons
(C) there is a dilation of the pupils owing to the stimulation of sympathetic neurons
(D) there is a response that can be blocked by atropine
(E) there is a contraction of the medial rectus muscle of the eye

686. A subject is given a cycloplegic (atropine) and asked to read a chart 20 feet away. He complains of blurred vision. The physician, by use of corrective lenses, concludes the image is blurred because it is being focused not on the retina but behind it. What conclusions can you draw from these observations? The patient is suffering from

(A) myopia
(B) hyperopia
(C) presbyopia
(D) either myopia or hyperopia
(E) either myopia or presbyopia

687. A bus driver complains of headaches and eye strain after each day's work. She is given a pair of glasses that have a prescription for each lens of +3 diopters (spherical) and is told to wear them while driving the bus. The results are that her headaches and eye strain disappear. This patient is most likely suffering from

(A) myopia
(B) hyperopia
(C) emmetropia
(D) astigmatism
(E) strabismus

688. What optical defect in the eye most commonly necessitates the wearing of bifocal lenses?

(A) myopia
(B) hyperopia
(C) presbyopia
(D) astigmatism
(E) strabism

689. Which one of the following statements is INCORRECT? The fovea centralis is an area in the retina in which there

(A) is the greatest concentration of cones
(B) is the greatest visual acuity
(C) are no rods
(D) is a greater sensitivity to light than any other area in the retina
(E) are no blood vessels overlying the light receptors

690. Which of the following statements is INCORRECT?

(A) the blind spot or optic disk in the retina is an area where the optic nerve leaves and blood vessels enter the eye
(B) the rods and cones are apposed to the sclera of the eye
(C) light, in passing to the rods and cones, must pass through a layer of ganglion cells and a layer of bipolar cells
(D) the macula lutea is a yellowish spot at the posterior pole of the eye that contains the fovea centralis
(E) the iris and ciliary muscles are smooth muscle that constitute the intrinsic muscles of the eye

691. Which of the following statements is INCORRECT?

(A) the near point is the closest distance an object can be brought to the eye and still be seen distinctly
(B) in aging, the near point gets closer to the eye beginning at age 20 or before
(C) dark adaptation takes about 20 minutes or longer
(D) wearing red goggles for a half hour before going out into the dark gives a more rapid dark adaptation than wearing blue goggles
(E) rods are more likely to become nonfunctional in bright light than cones

692. When light enters one eye

(A) the ipsi- and contralateral pupils constrict
(B) there is an ipsilateral reflex stimulation of sympathetic neurons to the iris
(C) there is a contralateral reflex stimulation of sympathetic neurons to the iris
(D) there will be a change in the size of the pupil that will occur even if the second cranial nerve has been sectioned bilaterally
(E) there will be a change in the size of the pupil that will occur even if the third cranial nerve has been sectioned bilaterally

693. Which of the following is INCORRECT? Light rays from an object to the left of the visual axis (ie, from the left temporal field)

(A) form an inverted, reversed image on the side of each retina to the right of the fovea centralis
(B) generate impulses that are carried in the right optic tract

(C) stimulate neurons which synapse in the right lateral geniculate nucleus
(D) stimulate neurons which synapse in the right occipital cortex
(E) generate impulses which all cross over in the optic chiasma

694. A patient with a total transection of the right optic tract would have which of the following symptoms?

(A) an ipsilateral loss of the nasal field and a contralateral loss of the temporal field of vision
(B) an ipsilateral and contralateral loss of the nasal fields
(C) an ipsilateral and contralateral loss of the temporal fields
(D) an ipsilateral loss of the temporal field and a contralateral loss of the nasal field
(E) an ipsilateral loss of vision

695. A patient with a total transection of the optic chiasma would have which of the following symptoms?

(A) a loss of the nasal field of the right eye and a loss of the temporal field of vision of the left eye
(B) a loss of the nasal fields of both eyes
(C) a loss of the temporal fields of both eyes
(D) a loss of the temporal field of the right eye and a loss of the nasal field of the left eye
(E) a loss of vision in the right eye

696. A patient with a total transection of the right optic nerve would have which of the following symptoms?

(A) an ipsilateral loss of the nasal field and a contralateral loss of the temporal field of vision
(B) an ipsilateral and contralateral loss of the nasal fields
(C) an ipsilateral and contralateral loss of the temporal fields
(D) an ipsilateral loss of the temporal field and a contralateral loss of the nasal field
(E) an ipsilateral loss of vision

697. Which of the following statements is INCORRECT?

(A) miosis is produced by narcotics such as morphine
(B) mydriasis is produced by parasympatholytic agents such as atropine
(C) mydriasis is produced by brain ischemia
(D) miosis is produced by near accommodation
(E) miosis is produced by anxiety

698. A patient is found to have lost the light reflex (miosis in response to light) but to have otherwise normal vision. He can, for example, see distinctly both near and distant objects. What lesion might explain this set of symptoms?

(A) a transection through the geniculocalcarine tract
(B) a transection through the superior part of the sympathetic chain
(C) a transection through the third cranial nerve
(D) a lesion in the tectal region of the midbrain
(E) a lesion in the occipital cortex

Answers and Explanations

683. (E)

$$\text{Refractive power} = \frac{1}{\text{focal length in meters}}$$

$$+\frac{1}{0.02\ \text{m}} = 50\ \text{diopters}$$

684. (D) Light rays from a single point 20 or more feet away strike the cornea essentially parallel to one another and in an emmetropic individual come to focus on a single point on the retina. Because of this relationship in optics objects 20 or more feet away are said to be at infinity. In an emmetropic individual objects closer than 20 feet away come to focus "behind the retina" (ie, they cause a blurred image). The response of an individual to a blurred image is called accommodation (decreased tension of suspensory ligaments).

685. (C) There is a constriction of the pupils associated with near accommodation. This causes a decreased field of vision when one is focused on a near object. Atropine prevents the contraction of the ciliary muscle and therefore prevents accommodation. Contraction of the medial rectus muscles causes convergence of the eyes.

686. (B) The patient is suffering from hyperopia. In myopia, the light from distant objects comes to focus in front of the retina. Presbyopia is a failure to accommodate due to a deterioration of the elastic characteristics of the lens.

687. (B) The bus driver is suffering from hyperopia. Her glasses (converging lenses) were prescribed to prevent eye strain due to the excessive use of the ciliary muscle, not to improve visual acuity. Myopia is corrected for by concave lenses (ie, lenses with a minus [–] diopter rating: diverging lenses).

688. (C) When there is a failure to accommodate (ie, presbyopia), bifocal or trifocal glasses are usually recommended. In this condition, the lens has lost much or all of its elastic recoil. Therefore, when the ciliary muscle contracts and the suspensory ligament decreases its tension on the lens, the lens *does not* change its convexity. In other words, there is either an inadequate or no accommodation. The major cause of presbyopia is aging. During the aging process, there is a progressive loss of elasticity throughout the body, ie, in the lens of the eye (presbyopia), in the lungs (emphysema), in the arteries (arteriosclerosis), and in the skin (wrinkles and bags under the eyes).

689. (D) Areas peripheral to the fovea centralis are more sensitive to light than the fovea centralis. The high concentration of cones and the fact that each sensory unit consists of but one cone contributes to the great visual acuity in this area.

690. (B) The rods and cones are apposed to the choroid layer of the eye, not the sclera.

691. (B) Starting in the teens, the near point begins to recede (ie, the ability of the lens to increase its convexity decreases). At age 10, the average near point is about 8 cm, and at age 70, it is about 100 cm (ie, over 3 feet). Because the rods are less sensitive to red light than to yellow, blue, or violet light, one can permit the rods to dark adapt while using cone vision to work in a moderate intensity of red light.

692. (A) In human beings, unlike some animals, light in one eye causes a reflex miosis in both eyes. There is a reflex stimulation of parasympathetic neurons and a depression of sympathetic neurons to the iris. Both the oculomotor and the optic nerve are essential for this reflex.

693. (E) Only the neurons from the left eye would carry these signals through the optic chiasma to the contralateral side. Neurons stimulated in the right eye would not cross over in the chiasma.

694. (A) The right optic tract contains nerve fibers from the lateral part of the ipsilateral retina (stimulated by light from the nasal field) and from the medial part of the contralateral retina (stimulated by light from the temporal field).

695. (C) The optic chiasm contains nerve fibers from the medial parts of the retina (stimulated by light from the temporal fields of vision).

696. (E) The right optic nerve carries fibers from the right eye.

697. (E) Anxiety activates sympathetic neurons and leads to mydriasis (large pupils).

698. (D) The phenomenon described here is called an Argyll-Robertson pupil. It is seen in neurosyphilis. Damage to the geniculocalcarine tract will cause a partial blindness and not eliminate the light reflex. Since, in the light reflex, miosis is brought about by a reflex stimulation of parasympathetic neurons, destruction of sympathetic neurons will not eliminate the light reflex. Damage to the third cranial nerve will eliminate the light reflex but will also eliminate accommodation to a near object.

CHAPTER 32

Other Senses

Questions

DIRECTIONS (Questions 699 through 715): Each of the numbered items or incomplete statements in this section is followed by answers or by completions of the statement. Select the ONE lettered answer or completion that is BEST in each case.

699. The threshold of audibility for a normal young adult is about 2×10^{-4} dynes/cm^2 and occurs in the middle of the audible range (ie, at about 2000 Hz). If a sound at this frequency produced 1000 times this pressure, it would have an intensity rating of

(A) less than 50 decibels (db)
(B) between 50 and 70 db
(C) between 70 and 90 db
(D) between 90 and 110 db
(E) greater than 110 db

700. Which of the following statements is correct?

(A) obstruction of the auditory tube will, in time, usually cause an outward bulging of the tympanic membrane
(B) the tympanic membrane continues vibrating several seconds after sound waves stop
(C) the tensor tympani and stapedius are skeletal muscles inserted on the malleus and stapes that contract in response to loud sounds
(D) the tectorial membrane separates the scala vestibuli from the scala media
(E) high-pitched sounds generate waves with a maximum amplitude near the apex of the cochlea, whereas low-pitched sounds produce their maximum amplitude near the base of the cochlea

701. Which of the following statements is correct?

(A) sound waves with an intensity of 0 decibels are inaudible
(B) loud sounds cause a reflex dampening of the movement of the auditory ossicles
(C) the ossicles are the only way of conducting sound waves across the middle ear
(D) the helicotrema is at the base of the cochlea
(E) deafness due to nerve damage is more improved by a hearing aid than deafness due to ossicular damage

702. Two patients are tested as follows. A tuning fork is set into vibration and placed on the mastoid process. Subject 1 reports he can initially hear the tuning fork distinctly but when its sound disappears, if it is held in the air next to his ear, he hears it again. Subject 2 reports she can hear the tuning fork only when it is placed on her mastoid process. If you assume that subject 1 has normal hearing, what conclusions can you draw concerning subject 2?

(A) there is probably damage to her cochlear branch of the eighth cranial nerve
(B) there is probably damage to her organ of Corti
(C) there is probably damage to either her auditory nerve or the organ of Corti
(D) there is probably damage to either her tympanic membrane, ossicles, or oval window
(E) there is probably damage to her mastoid process

703. Which of the following statements is INCORRECT? Impulses in the right auditory nerve

(A) are initiated by distortion of hair cells in the organ of Corti
(B) are carried to the cochlear nuclei of the medulla (here first-order neurons synapse with second-order neurons)
(C) result in the stimulation of fourth-order neurons in the medial geniculate body of the thalamus
(D) result in the stimulation of the occipital cortex (auditory area)
(E) strongly stimulate both the ipsilateral and contralateral auditory areas of the cortex

704. Endolymph

(A) lies in the bony labyrinth
(B) has Na^+ and K^+ concentration resembling extracellular fluid
(C) has a Na^+ concentration higher than extracellular fluid
(D) bathes the organ of Corti and the cupulae of the semicircular canals
(E) when one is traveling at a constant velocity produces a displacement of the cupula of the semicircular canal

705. Which of the following statements is INCORRECT?

(A) the semicircular canals do not lose their function under weightless conditions
(B) when cold water is infused over the eardrum, nystagmus results
(C) nystagmus may result from damage to the semicircular canals or cerebellum
(D) nystagmus is a quick movement of the eyes and is preceded by a slow movement
(E) nystagmus may be caused by damage to the cochlea, retina, or one of the first six cranial nerves

706. Which of the following sensations is served by sensory neurons with no primary relay in the thalamus and no projection area on the neocortex?

(A) touch
(B) acceleration
(C) taste
(D) smell
(E) taste and smell

707. Which of the following statements is INCORRECT?

(A) the receptor neuron in the olfactory epithelium synapses in the olfactory bulb
(B) some of the axons in the olfactory tract pass via the anterior commissure to the contralateral olfactory bulb
(C) the rhinencephalon is almost totally concerned with olfactory functions
(D) the rhinencephalon is also called the limbic system
(E) the odors of peppermint, menthol, and chlorine stimulate neurons in the trigeminal nerve

708. Which of the following statements is INCORRECT?

(A) the taste buds are confined to the tongue
(B) the sensory neurons serving the sense of taste are in the seventh, ninth, and tenth cranial nerves
(C) first-order neurons from the taste buds synapse in the nucleus solitarius
(D) second-order neurons from the nucleus solitarius send axons up the contralateral medial lemniscus
(E) if all of the sensory neurons from a taste bud are destroyed, the taste bud will degenerate

709. During an operation on the brain, a surgeon exposes a part of the lateral surface of the postcentral gyrus halfway between the medial longitudinal fissure and the temporal lobe. She then stimulates this area with fine electrodes and asks the patient what he feels. The patient

(A) might reply that he feels as if something were touching his index finger
(B) might reply that he feels heat in his foot
(C) might reply that he sees light
(D) might reply that he feels a tingling in his head
(E) would not reply because brain surgery cannot be performed on a conscious subject

710. A patient is asked to stand erect with his feet close together. With his eyes open, he seems quite normal, but when he closes his eyes, there is a pronounced body sway, and the patient sometimes loses his balance. Other signs of malfunction are an elevated touch threshold, an impaired ability to localize sensation, and an impaired ability to recognize objects such as keys by touching them (astereognosis). Which one of the following is most likely to be the cause of this condition?

(A) bilateral destruction of the ventral spinothalamic tract
(B) bilateral destruction of the dorsal columns
(C) a cerebellar lesion
(D) bilateral destruction of the corticospinal tract
(E) a transection of the midbrain

711. All of the following are symptoms of a bilateral loss of the dorsal columns EXCEPT

(A) ataxia: a loss of muscle coordination and a resultant abnormal gait.
(B) inability to recognize limb position with eyes closed.
(C) loss of two-point discrimination: cannot distinguish between a single touch stimulus and two stimuli applied simultaneously.
(D) loss of vibratory sense: patient cannot distinguish between a vibrating tuning fork and a silent one by touching them.
(E) loss of temperature sense

712. Irritation of the peritoneum covering the diaphragm will produce a pain that is usually referred to the

(A) testes
(B) ventral surface of the abdomen
(C) skin between the 11th and 12th rib
(D) shoulder or neck
(E) head

713. Which of the following statements is INCORRECT? Enkephalins

(A) decrease intestinal motility
(B) increase the release of substance P by first-order pain fibers
(C) compete with morphine for receptor sites
(D) function as synaptic transmitters
(E) depress pain perception, produce a state of euphoria, and cause pupillary constriction

714. Lesions in the central nervous system are sometimes produced surgically to treat intractable pain or may develop from a pathologic process. Some of the following are procedures or processes that affect the sensation of pain. Which of the following statements is INCORRECT?

(A) prefrontal lobotomy: the white matter of the frontal lobe is incised; a markedly increased threshold to the sensation of pain results

(B) unilateral anterolateral cordotomy: a lateral spinothalamic tract is incised; a loss of the pain and temperature sense in the skin of the contralateral side below the lesion results

(C) unilateral lesion in the postcentral gyrus: little change in the threshold for pain on either side of the body results

(D) syringomyelia in the cervical enlargement of the cord: destruction of fibers crossing the cord below the central canal; a loss of the pain and temperature sense restricted to the upper extremities results

(E) tic douloureux: attacks of severe pain in the face, lips, or tongue due to stimulation of the trigeminal nerve

715. Which of the following statements is INCORRECT?

(A) proprioception and fine touch are affected by a cortical lesion to approximately the same degree as pain sensation

(B) the size of the cortical receiving area on the postcentral gyrus for a part of the body is proportionate to the number of receptors in that part

(C) an increase in the frequency of impulses carried by a neuron or an increase in number of neurons stimulated are both signals to the central nervous system that are interpreted as an increase in intensity of sensation

(D) an individual who, because of disease or heredity, has no functional, myelinated pain fibers can still feel pain but will have a slower reflex withdrawal in response to intense heat than a normal subject

(E) most visceral sensation travels the same pathway as somatic sensation in the spinothalamic tracts and the thalamic radiations

Answers and Explanations

699. (B)

$$\text{Intensity in db} = 20 \cdot \left(\log \frac{\text{measured pressure}}{\text{standard pressure}}\right)$$

$$= 20 \cdot (\log 1000) = 20 \cdot 3 = 60\text{db}$$

700. (C) The attenuation reflex (contraction of tensor tympani and stapedius muscles) can reduce the intensity of low-frequency sound by 30 to 40 decibels. This protects the auditory receptors from excessive stimulation. Low-pitched sounds generate their maximum amplitude wave near the apex of the cochlea.

701. (B) This attenuation reflex protects the cochlea from damage. Zero decibels represents a sound intensity of 2×10^{-4} dynes/cm^2. Most normal, young adults can hear a frequency of 2000 Hz at this intensity. Loud sounds can set up vibrations through either the oval or round window in the absence of a functional tympanic membrane or functional ossicles. The helicotrema is at the apex of the cochlea and connects the perilymph of the scala vestibuli with that of the scala tympani. Because of this path, a bulging of the oval window inward will cause a bulging of the round window outward. Hearing is impossible in the absence of the auditory nerves. A hearing aid merely amplifies the signal delivered to the organ of Corti.

702. (D) This is the Rinne test: The tuning fork held against the mastoid process is delivering vibrations to the perilymph, endolymph, and organ of Corti, which, in turn, stimulates neurons in the auditory nerve. In short, the major defect here is apparently in the delivery of impulses to the inner ear. The middle ear delivers impulses to the inner ear. Occlusion of the auditory tube can produce sufficient impairment of tympanic function to produce the above symptoms, as can otitis media (impaired function of the ossicles) and otosclerosis (immobilization of the stapes and oval window).

703. (D) The auditory area of the cortex is the superior temporal gyrus.

704. (D) Endolymph lies in the membranous labyrinth and bathes the organ of Corti and the cupulae of the semicircular canal. Its concentration of Na^+ and K^+ resembles that of intracellular fluid, ie, is rich in K^+. It produces a displacement of the cupulae during acceleration and deceleration but not during a constant velocity.

705. (E) None of these systems when damaged produce nystagmus. The semicircular canals continue to be stimulated by acceleration and deceleration under weightless conditions. The caloric test is used clinically to assess unilateral semicircular canal function.

706. **(D)** Impulses pass from the olfactory epithelium to the prepyriform cortex and the periamygdaloid area without passing through the thalamus. From these areas, impulses may then be relayed to the thalamus and hypothalamus.

707. **(C)** In human beings, only a small part of the rhinencephalon is concerned with olfactory function. It is much more concerned with sexual behavior, the emotions, and motivation. Since it is realized that only a small part of this system in human beings is concerned with olfaction, it is more common to call it the limbic system than the "smell brain." Nasal irritants like peppermint act by stimulating free nerve endings that travel with the fifth cranial nerve. These endings may also initiate sneezing, lacrimation, and respiratory inhibition.

708. **(A)** Taste buds are found on the tongue, epiglottis, and pharynx. The facial and glossopharyngeal nerves innervate the tongue and the vagus innervates the epiglottis and pharynx.

709. **(A)** Sensations for the hand and finger are located in this area. Other sensations that might also be felt are numbness, tingling, movement, warmth, cold, and pain. The area for the foot is adjacent to the superior sagittal sinus. The postcentral gyrus is in the parietal lobe of the cortex. Sensations of light are caused by stimulation of the occipital lobe. Although the stimulus is being applied to the brain the sensation is being projected to the finger in much the same way a blind person projects his sense of feeling to the tip of his cane. Brain surgery on a conscious subject is sometimes necessary when the patient's response to the stimulation of certain areas in the brain is needed. In this technique, sensation in the scalp is eliminated by a local anesthetic. Since there are no pain fibers in the brain (just internuncial neurons), this procedure can be relatively painless.

710. **(B)** The dorsal columns are more important than other ascending tracts in the detailed localization of touch sensation, the determination of form, and in balance with the eyes closed. The ventral spinothalamic tract carries touch and pressure fibers, and its destruction will cause an elevated touch threshold but will not produce the other symptoms listed. In both a cerebellar lesion that produced problems with balance and a midbrain transection, the balance problem would exist with both eyes open and closed. Bilateral destruction of the corticospinal tract would result in loss of voluntary control of skeletal muscle. Stereognosis would be normal.

711. **(E)** Temperature fibers travel in the spinothalamic tract.

712. **(D)** When pain is referred, it is usually projected to a structure that developed from the same dermatome (ie, the same embryonic segment) as the irritated structure. Thus, pain caused by an irritation (1) of the diaphragm is referred to the neck or shoulder, (2) of the heart to the chest, axilla, or inside of the left arm, (3) of the ureter to the groin or testes, (4) of the stomach to between the scapulae, and (5) of the appendix to the umbilical area.

713. **(B)** Enkephalins apparently act by presynaptic inhibition to decrease the release of substance P by first-order pain fibers. It has been suggested that acupuncture has its analgesic effect by causing these opioid peptides to be released. They, like morphine, may have the effects listed in statement E.

714. **(A)** Prefrontal lobotomy has little effect on the threshold to pain. It does cause a marked change in attitudes, however. After this operation, pain usually is no longer an unpleasant sensation. Bilateral anterolateral cordotomy is a neurosurgical procedure for the relief of pain in terminal cancer patients. It does not produce the marked personality changes that result from lobotomy. Lesions in the postcentral gyrus may cause agnosias (inability to recognize certain stimuli), loss of fine touch discrimination, and disorders of spatial discrimination, but they produce only mild changes in the threshold for pain. Pain apparently can be recognized at the subcortical level.

715. (A) Proprioception and fine touch are more affected by cortical lesions than pain. Unmyelinated and myelinated fibers carry the sensation of pain. The former conduct more slowly than the latter. A child lacking myelinated pain fibers is much more likely to be injured than a normal child.

CHAPTER 33

Central Nervous System

Questions

DIRECTIONS (Questions 716 through 739): Each of the numbered items or incomplete statements in this section is followed by answers or by completions of the statement. Select the ONE lettered answer or completion that is BEST in each case.

716. All of the following structures are found in the medulla oblongata EXCEPT

(A) hypoglossal nucleus
(B) pyramidal decussation
(C) red nucleus
(D) medial and spinal vestibular nuclei
(E) reticular formation

717. All of the following structures are found in the pons EXCEPT

(A) motor nucleus of the facial nerve
(B) nucleus of the abducent nerve
(C) lateral geniculate body
(D) lateral and superior vestibular nuclei
(E) reticular formation

718. All of the following structures are found in the midbrain EXCEPT

(A) substantia nigra
(B) fornix
(C) corpora quadrigemina
(D) red nucleus
(E) reticular formation

719. All of the following structures are found in the diencephalon EXCEPT

(A) mamillary bodies
(B) pineal body
(C) subthalamus
(D) lateral geniculate bodies
(E) reticular formation

720. A unilateral destruction of extrapyramidal fibers in the medulla causes

(A) spastic paralysis contralateral to the lesion
(B) muscle fasiculations
(C) hyporeflexia
(D) ipsilateral hypotonicity
(E) flexion of the leg

Questions 721 and 722

A patient presents the following history: One month ago, he had begun to experience pain in the ventral surface of the right thigh and leg. On examination, the physician noted an impaired vibratory sense and tactile discrimination in the leg. There was no Babinski sign on either side. The upper extremities and the contralateral leg appeared normal.

721. These findings are most consistent with a tumor

(A) at S4
(B) at T12
(C) at C2
(D) compressing the dorsal root
(E) compressing the ventral root

722. If the signs in this case are caused by a tumor, the tumor apparently has disrupted the function of the

(A) lateral corticospinal tract
(B) the ipsilateral dorsal column
(C) the contralateral dorsal column
(D) the lateral spinothalamic tract
(E) the dorsal column and the corticospinal tract

Questions 723 through 726

A patient had the following clinical signs:

1. motor paralysis, hypertonicity, and hyperreflexia on the left side of the body below the neck with the exception of the diaphragm, shoulders, and parts of the arm, forearm, and hand
2. loss of the position sense, tactile discrimination sense, and vibratory sense in the same areas
3. numbness in the left thumb that was noted nowhere else in the body
4. loss of pain and temperature sense below the neck on the right side of the body with the exception of the shoulders and parts of the arm, forearm, and hand

723. The cause of these signs is probably a lesion in

(A) the left side of the pons
(B) the right side of the pons
(C) the left side of the spinal cord
(D) the right side of the spinal cord
(E) none of the above areas

724. The clinical signs in this case are consistent with a diagnosis of a lesion restricted to the left

(A) dorsal and lateral funiculi
(B) dorsal funiculus
(C) ventral funiculus
(D) lateral and ventral funiculi
(E) rubrospinal tract

725. The clinical signs in this case are consistent with a diagnosis of a hemisection of the spinal cord. It probably occurred at

(A) C2
(B) C6
(C) T2
(D) T5
(E) L1

726. Why, in this case, is there a numbness of the left thumb but no complete loss of the sensation of touch anywhere else? The lesion

(A) destroyed the left dorsal root but left intact the right ventral spinothalamic tract
(B) destroyed the left dorsal root but left intact the right dorsal funiculus
(C) destroyed the ventral horn but left intact the right ventral spinothalamic tract
(D) destroyed the ventral horn but left intact the right dorsal funiculus
(E) left intact the right lateral spinothalamic tract

727. A unilateral transection of the spinal cord causes below the lesion

(A) an ipsilateral loss of the vibration sense and tactile discrimination and a contralateral increase in the touch threshold
(B) an ipsilateral loss of the vibration sense and tactile discrimination and a contralateral decrease in the touch threshold
(C) a contralateral loss of the vibration sense and tactile discrimination and a contralateral decrease in the touch threshold
(D) a contralateral loss of the vibration sense and tactile discrimination and a contralateral increase in the touch threshold
(E) a contralateral spastic paralysis

728. A patient who is naturally righthanded has his corpus callosum severed. In a clinical test, he is shown a series of pictures and told to say the word "horse" every time he sees a picture of a horse. When both of his eyes are open, he performs this task without error. If his optic chiasma is

(A) also severed, he probably could not perform this task if his right eye is closed
(B) also severed, he probably could not perform this task if his left eye is closed

(C) also severed, he probably could not perform this task with either eye closed
(D) either intact or severed, he probably could not perform this task if his right eye is closed
(E) either intact or severed, he probably could not perform this task if his left eye is closed

729. The corpus callosum has been completely severed in several human subjects. Prior to this operation the subjects were suffering from frequent epileptic seizures that could not be controlled by drugs. After the operation, there was diminution in the frequency of the seizures. A number of tests were performed on these subjects. In one, the subject covers his left eye and sits at a table containing a number of objects (a key, a nut, a penny, etc). The visual field in the right eye that sends impulses to the left brain (ie, the right temporal visual field) is occluded. In other words, the subject is using only his right brain for vision. The subject is now told to pick up whatever object is named on a lantern slide. The word "nut" is shone on the screen. The subject will

(A) not pick up the nut
(B) pick up the nut with his right hand but be unable to tell you what he has done
(C) pick up the nut with his left hand but be unable to tell you what he has done
(D) pick up the nut with his right hand and be able to tell you what he has done
(E) pick up the nut with his left hand and be able to tell you what he has done

730. In a subject with the corpus callosum destroyed, the word "hat" was seen only by the right brain and the word "band" only by the left brain. If he were asked what he saw, he would say the word

(A) "band" and write with the left hand the word "hat"
(B) "hat" and write with the left hand the word "band"
(C) "band" and write with the left hand the word "band"
(D) "hat" and write with the left hand the word "hat"
(E) "hat" and write nothing with the left hand

731. The right hemisphere is better than the left hemisphere in its

(A) sensitivity to signals from the right cochlea
(B) sensitivity to signals from the left olfactory epithelium
(C) sensitivity to signals from the left cochlea and the left olfactory epithelium
(D) sensitivity to signals from the right cochlea and right olfactory epithelium
(E) nonverbal ideation (ie, constructing with colored blocks a mosaic that matches a colored picture, or most other types of copying)

732. Immediately following the transection of the spinal cord there is

(A) a period of spinal shock that rarely lasts more than 24 hours
(B) a general increase in skeletal muscle tone
(C) a retention of urine and feces
(D) a retention of urine and feces associated with an increase in skeletal muscle tone
(E) increased reflex emptying of the urinary bladder (automatic bladder)

733. A patient 3 weeks after a motor vehicle accident has (1) an increased tone in his antigravity muscles in the four limbs, (2) bilaterally hyperactive knee jerks, and (3) an inability to maintain his body temperature. In addition, it is noted that (4) muscle tone could be modified by turning the patient's head. The accident has apparently produced a

(A) hemidecortication
(B) decortication with an intact brain stem
(C) midcollicular transection
(D) bilateral section of the cord
(E) either a midcollicular transection or a bilateral section of the cord

734. A patient exhibits athetoid movements. She probably has a

(A) cerebellar lesion
(B) lesion in the basal ganglia
(C) thalamic lesion
(D) hypothalamic lesion
(E) lesion in the cerebral cortex

735. A child demonstrates irregular, spasmodic, involuntary movements of the limbs and facial muscles. He most likely has a lesion in the

(A) caudate nucleus
(B) precentral gyrus of the cortex
(C) postcentral gyrus of the cortex
(D) rubrospinal tract
(E) midline of the cerebellum

736. Which of the following substances is present in high concentration in the putamen and caudate nucleus and has been shown to be at a reduced concentration in Parkinson's disease?

(A) acetylcholine
(B) dopamine (3, 4-dihydroxyphenylethylamine)
(C) histamine
(D) GABA (gamma-aminobutyric acid)
(E) melatonin

737. The electroencephalogram (EEG) from a normal adult human who has her eyes closed and is letting her mind wander has as its predominant pattern

(A) an alpha rhythm (ie, a pattern with a lower frequency than a beta rhythm and a higher frequency than a theta or delta rhythm)
(B) a beta rhythm
(C) a delta rhythm
(D) a rhythm that is unaffected by changes in the blood chemistry
(E) a rhythm that is the same in the child under similar circumstances

738. Dreaming usually is associated with

(A) the period of sleep when the EEG contains slow waves (less than 8 Hz)
(B) rapid eye movements (REM)
(C) an increased muscle tone
(D) a decreased threshold to extrinsic stimuli
(E) sleep walking

739. The ascending, midbrain reticular formation

(A) is most active during wakefulness
(B) is most active when there is a minimal stimulation of the extero- and interoceptors of the body
(C) sends all or almost all of its impulses to the thalamus
(D) sends all or almost all of its impulses via the thalamus to the precentral gyrus
(E) has no known function in man

Answers and Explanations

716. **(C)** The red nucleus is located in the midbrain.

717. **(C)** The lateral geniculate body is located in the diencephalon.

718. **(B)** The fornix is located in the diencephalon.

719. **(E)** The reticular formation is located in the medulla oblongata, pons, and midbrain.

720. **(A)** Contralateral spasticity is due to the loss of inhibition of the stretch reflexes. Muscle fasciculations and hyporeflexia are characteristic of a lower motor neuron lesion. When there is hypertonicity, the leg is usually extended.

721. **(D)** The thighs, legs, and feet are innervated by neurons entering the cord at L2 through S2. The sensory nature of the problem suggests the involvement of the sensory root (ie, the dorsal root).

722. **(B)** The impaired vibratory sense and tactile discrimination are consistent with a diagnosis of ipsilateral dorsal column malfunction. The lack of a Babinski sign as well as other signs is consistent with the view that the tumor is having little or no effect on the lateral column. The dorsal column carries impulses from its ipsilateral side. The fibers in the lateral spinothalamic tract are stimulated by pain and temperature fibers from the contralateral side of the body and do not affect tactile discrimination and the vibratory sense.

723. **(C)** The lesion is on the same side as the motor paralysis and the proprioceptive deficit. The motor paralysis is caused by the destruction of descending fibers which have crossed over in the brain stem. The proprioceptive deficit is caused by the destruction of ascending fibers in the dorsal columns. They do not cross over in the spinal cord. The loss of the pain sense is contralateral to the lesion because pain fibers cross over almost as soon as they enter the cord. If a unilateral lesion were in the pons, clinical signs 1, 2, and 4 would be on the contralateral side. Since the loss of pain is on one side of the body and the motor paralysis on the other, a unilateral pontine lesion is unlikely.

724. **(A)** The clinical picture is the result of a destruction of the dorsal funiculus (sign 2) and lateral funiculus (sign 1 is due to the destruction of the lateral corticospinal tract; sign 4 is due to the loss of the lateral spinothalamic tract).

725. **(B)** A lesion at C6 leaves the innervation of the diaphragm and shoulder and parts of the arm and forearm intact. A lesion at C2 here would cause a motor paralysis of the diaphragm and an involvement of the scalp and neck, as well as all of the shoulder, arm, and forearm. A lesion at T2 would usually leave the upper extremities intact.

726. **(A)** Destruction of the left dorsal root at C6 usually causes a complete sensory loss in the left thumb. Touch fibers from the left side of the body (1) ascend the cord in the left dorsal funiculus (column) and (2) stimulate neurons which ascend the cord in the right ventral spinothalamic tract. Thus, a unilateral lesion in the cord will not cause a complete loss of the sensation of touch on either side. Destruction of the ventral horn will cause motor paralysis but no sensory deficit. The lateral spinothalamic tract carries pain and temperature signals but no touch signals.

727. **(A)** The loss of the vibration sense results from the destruction of the dorsal columns (uncrossed ascending axons), and the increased touch threshold results, at least in part, from the destruction of the ventral spinothalamic tract (crossed ascending axons).

728. **(B)** The left cerebral hemisphere is the part of the forebrain where most of the language functions are centered in right-handed individuals and over 70% in left-handed individuals. Since the destruction of the optic chiasma prevents impulses from passing from the retina of one eye to the contralateral cortex, and the destruction of the corpus callosum prevents most impulses from passing from one cortex to the contralateral brain, option (B) is the only possibility. Some of the other commissural tracts in the brain include the anterior commissure, the suprachiasmatic commissure, the posterior commissure, and the habenular commissure. They apparently cannot prevent the visual receptive aphasia noted in this case. Even if there is visual input into only one eye, that eye has connections via the optic chiasma to the contralateral cortex.

729. **(C)** He will not be able to tell you what he has done because an oral or written description requires sensory input into the left cerebrum. The right cerebrum is capable of simple language comprehension, but not the production of language. The left cerebrum plays an essential role in initiating motor activity in the right arm and leg in response to olfactory, visual, and auditory stimuli. Since the stimulus is being delivered to the right brain and the major connecting link between the right and left sensory areas has been destroyed, the right arm will not pick up the nut.

730. **(A)** If the subject were asked what kind of band it was, he would just as likely say "jazz band" as "hat band." Sperry received the Nobel Prize in Medicine in 1981 for much of his work on specialization of the hemispheres.

731. **(E)** The left hemisphere has been characterized as "almost illiterate in respect to pictorial and pattern sense, at least, as displayed by its copying disability." It has been suggested, for example, that the removal of the right temporal lobe seriously limits musical ability. There is an ipsilateral projection of smell and a predominantly contralateral projection of hearing.

732. **(C)** Spinal shock may last only 2 weeks or for several months. There is a decrease in muscle tone (flaccid paralysis) and a loss of reflexes during the period of spinal shock. The tendency to retain urine can result in a genitourinary tract infection.

733. **(C)** The signs of this lesion are decerebrate rigidity (extension of the arms and legs). The temperature regulation problem is due to the lesion separating most of the body from descending hypothalamic neurons. In decortication, you obtain extension of the hind limbs but a moderate flexion of the arms. This is called decorticate rigidity.

734. **(B)** Lesions in the basal ganglia produce involuntary contractions in the otherwise inactive individual. Athetosis may be produced by a lesion in the lenticular nucleus. In athetosis, there is a continuous series of writhing, worm-like movements in the extremities. In a cerebellar lesion, you will generally find few problems in the resting individual. Here the major problems concern walking, grabbing, talking, and other types of willed activity.

735. **(A)** Hyperkinetic syndromes such as the one discussed above (chorea or St. Vitus' dance) are usually caused by a malfunction of the basal ganglia. Lesions of the rubrospinal tract cause spasticity. Lesions of the cerebellum cause difficulty in maintaining an upright stance and are associated with a staggering gait.

736. **(B)** In parkinsonism the dopaminergic neurons in the substantia nigra are functionally impaired causing a loss of inhibition to the caudate nucleus and putamen. The administration of dopamine has proven successful in the treatment of parkinsonism.

737. **(A)** During the alpha rhythm the wave frequency is 8 to 12 Hz. Decreases in blood sugar, glucocorticoids, body temperature, or increases in arterial PO_2 decrease the frequency of alpha waves. The predominant pattern prior to adolescence is the theta rhythm (4 to 7 Hz).

738. **(B)** Rapid eye movements (REM sleep) are associated with dreaming. During this phase the EEG patterns are similar to those in an alert subject (high frequency, low voltage). There is a decreased muscle tone, and arousal is difficult. Sleep walking occurs during slow wave sleep (stage 4).

739. **(A)** Most of the neurons carrying sensory messages to the cortex give off collaterals, which stimulate the reticular-activating system (RAS) in the reticular formation. As a result, many signals are sent from the RAS that bypass the thalamus and project diffusely to the cortex. These signals serve to keep the subject alert.

APPLETON & LANGE REVIEW SERIES

HEALTH RELATED

Appleton & Lange's Review of Cardiovascular-Interventional Technology
Vitanza
1995, ISBN 0-8385-0248-2, A0248-3

Appleton & Lange's Review for the Chiropractic National Boards, Part I
Shanks
1992, ISBN 0-8385-0224-5, A0224-4

Appleton & Lange's Review for the Dental Assistant, 3/e
Andujo
1992, ISBN 0-8385-0135-4, A0135-2

Appleton & Lange's Review for the Dental Hygiene National Board Review, 5/e
Barnes and Waring
1998, ISBN 0-8385-0342-X, A0342-4

Appleton & Lange's Review for the Medical Assistant, 5/e
Palko and Palko
1997, ISBN 0-8385-0285-7, A0285-5

Appleton & Lange's Review of Pharmacy, 6/e
Hall and Reiss
1997, ISBN 0-8385-0281-4, A0281-4

Appleton & Lange's Review for the Physician Assistant, 3/e
Cafferty
1997, ISBN 0-8385-0279-2, A0279-8

Appleton & Lange's Review of Physiology for the USMLE Step 1
Penney
1998, ISBN 0-8385-0274-1, A0274-9

Appleton & Lange's Review for the Radiography Examination, 3/e
Saia
1997, ISBN 0-8385-0280-6, A0280-6

Appleton & Lange's Review for the Surgical Technology Examination, 4/e
Allmers and Verderame
1996, ISBN 0-8385-0270-9, A0270-7

Appleton & Lange's Review for the Ultrasonography Examination, 2/e
Odwin
1993, ISBN 0-8385-9073-X, A9073-6

Essentials of Advanced Cardiac Life Support: Program Review & Exam Preparation (PREP)
Brainard
1997, ISBN 0-8385-0259-8, A0259-0

Radiography: Program Review & Exam Preparation (PREP)
Saia
1996, ISBN 0-8385-8244-3, A8244-4

FIRST AID

1998 First Aid for the USMLE Step 1
A Student-to-Student Guide
Bhushan, Le, and Amin
1998, ISBN 0-8385-2603-9, A2603-7

First Aid for the USMLE Step 2, 2/e
A Student-to-Student Guide
Go, Curet-Salim, and Fullerton
1998, ISBN 0-8385-2604-7, A2604-5

First Aid for the Wards
A Student-to-Student Guide
Le, Bhushan, and Amin
1997, ISBN 0-8385-2594-4, A2595-5

First Aid for the Match
A Student-to-Student Guide
Le, Bhushan, and Amin
1996, ISBN 0-8385-2596-2, A2596-3

FLASH FACTS

Flash Facts for USMLE Steps 2 and 3
Kaiser and Kaiser
1998, ISBN 0-8385-2606-3, A2606-0

INSTANT EXAM

The Instant Exam Review for the USMLE Step 2, 2/e
Goldberg
1996, ISBN 0-8385-4328-6, A4328-9

The Instant Exam Review for the USMLE Step 3, 2/e
Goldberg
1997, ISBN 0-8385-4337-5, A4337-0

KAPLAN®

USMLE Step 1 Starter Kit
Kaplan® and Appleton & Lange
1998, ISBN 0-8385-8665-1, A8665-0

USMLE Step 2 Starter Kit
Kaplan® and Appleton & Lange
1998, ISBN 0-8385-8666-X, A8666-8

COMPREHENSIVE A&L REVIEWS

Appleton & Lange's Review for the USMLE Step 1, 2/e
Barton
1996, ISBN 0-8385-0265-2, A0265-7

Appleton & Lange's Review for the USMLE Step 2, 2/e
Catlin
1996, ISBN 0-8385-0266-0, A0266-5

Appleton & Lange's Review for the USMLE Step 3, 2/e
Jacobs
1997, ISBN 0-8385-0305-5, A0305-1

BASIC SCIENCE

Appleton & Lange's Review of Anatomy for the USMLE Step 1, 5/e
Montgomery
1995, ISBN 0-8385-0246-6, A0246-7

Appleton & Lange's Review of Epidemiology & Biostatistics for the USMLE
Hanrahan and Madupu
1994, ISBN 0-8385-0244-X, A0244-2

Appleton & Lange's Review of Microbiology and Immunology, 3/e
Yotis
1996, ISBN 0-8385-0273-3, A0273-1

Appleton & Lange's Review of General Pathology, 3/e
Lewis and Barton
1993, ISBN 0-8385-0161-3, A0161-8

CLINICAL SCIENCE

Appleton & Lange's Review of Internal Medicine
Goldlist
1996, ISBN 0-8385-0251-2, A0251-7

Appleton & Lange's Review of Obstetrics and Gynecology, 5/e
Julian, et al.
1995, ISBN 0-8385-0231-8, A0231-9

Appleton & Lange's Review of Pediatrics, 6/e
Lorin
1998, ISBN 0-8385-0303-9, A0303-6

Appleton & Lange's Review of Psychiatry, 5/e
Easson
1994, ISBN 0-8385-0247-4, A0247-5

Appleton & Lange's Review of Surgery, 3/e
Wapnick
1998, ISBN 0-8385-0245-8, A0245-9

SPECIALTY BOARD REVIEWS

The MGH Board Review of Anesthesiology, 4/e
Dershwitz
1994, ISBN 0-8385-8611-2, A8611-4

More on reverse →